ASCE Manuals and Reports on Engineering Practice No. 141

Wood Pole Structures for Electrical Transmission Lines

Recommended Practice for Design and Use

Prepared by the
Task Committee on Recommended Practice for
Wood Pole Structures for Electrical Transmission Lines of the
Structural Engineering Institute of the
American Society of Civil Engineers

Edited by
James M. McGuire, P.E.,
Otto J. Lynch, P.E.,
Vicki Schneider, P.E.,
Jaron T. Reay, P.E., S.E.,
Timothy P. Wachholz, P.E., S.E.,
and Nelson G. Bingel III

ASCE AMERICAN SOCIETY OF CIVIL ENGINEERS

Published by the American Society of Civil Engineers

Library of Congress Cataloging-in-Publication Data

Names: Structural Engineering Institute. Task Committee on Recommended Practice for Wood Pole Structures for Electrical Transmission Lines, author. | McGuire, James M., 1963– editor. | Lynch, Otto J., editor. | Schneider, Vicki, editor.
Title: Wood pole structures for electrical transmission lines: recommended practice for design and use / prepared by the Task Committee on Recommended Practice for Wood Pole Structures for Electrical Transmission Lines of the Structural Engineering Institute of the American Society of Civil Engineers; edited by James M. McGuire, P.E., Otto J. Lynch, P.E. and Vicki Schneider, P.E.
Description: Reston, Virginia: American Society of Civil Engineers, [2019] | Includes bibliographical references and index.
Identifiers: LCCN 2018056536 | ISBN 9780784415245 (soft cover: alk. paper) | ISBN 9780784415245 (print) | ISBN 9780784448205 (PDF)
Subjects: LCSH: Electric lines—United States—Poles and towers—Design and construction. | Electric lines—United States—Poles and towers—Installation. | Wood poles.
Classification: LCC TK3242 .S77 2019 | DDC 621.319/22—dc23
LC record available at https://lccn.loc.gov/2018056536

Published by American Society of Civil Engineers
1801 Alexander Bell Drive
Reston, Virginia 20191-4382
www.asce.org/bookstore | ascelibrary.org

***Errata:* Errata, if any, can be found at https://doi.org/10.1061/9780784415245.**

ISBN 978-0-7844-1524-5 (print)
ISBN 978-0-7844-4820-5 (PDF)
Manufactured in the United States of America.

25 24 23 22 21 20 19 1 2 3 4 5

Cover photo credits:
Front: Upper left and right: Courtesy of Otto J. Lynch, P.E.
Front: Lower left: Great River Energy, used with permission
Back: Courtesy of Otto J. Lynch, P.E.

MANUALS AND REPORTS ON ENGINEERING PRACTICE

(As developed by the ASCE Technical Procedures Committee, July 1930, and revised March 1935, February 1962, and April 1982)

A manual or report in this series consists of an orderly presentation of facts on a particular subject, supplemented by an analysis of limitations and applications of these facts. It contains information useful to engineers in their everyday work, rather than findings that may be useful only occasionally or rarely. It is not in any sense a "standard," however; nor is it so elementary or so conclusive as to provide a "rule of thumb" for nonengineers.

Furthermore, unlike a paper (which expresses only one person's observations or opinions), the material in this series is the work of a committee or group selected to assemble and express information on a specific topic. As often as practicable, the committee is under the direction of one or more Technical Divisions and Councils, and the product has evolved and been subjected to review by the Executive Committee of the Division or Council. As a step in this review process, proposed manuscripts are often brought before the members of the Technical Divisions and Councils for comment, which may serve as the basis for improvement. When published, each work shows the names of the committees that compiled the work and indicates clearly the processes through which it has passed in review so that its merit may be definitely understood.

In February 1962 (and revised in April 1982), the Board of Direction voted to establish a series titled "Manuals and Reports on Engineering Practice," which includes the manuals published and authorized to date, future Manuals of Professional Practice, and Reports on Engineering Practice. All such manuals or report materials of the Society would have been refereed in a manner approved by the Board Committee on Publications and would be bound, with applicable discussion, in books similar to past manuals. Numbering would be consecutive and would be a continuation of present manual numbers. In some cases of joint committee reports, the bypassing of journal publications may be authorized.

A list of available Manuals of Practice can be found at http://www.asce.org/bookstore.

CONTENTS

PREFACE

The purpose of this manual is to provide the reader with a basic knowledge of the principles and methods for the design and use of wood poles for overhead utility line structures. This manual is the result of a multiyear collaborative effort by engineers and technical experts from electric utilities, consulting firms, academia, and manufacturers engaged in the design and application of these structures.

Every effort has been made to ensure the accuracy and clarity of this document. The user is reminded that utility line structures are an integral part of a larger overhead line system and therefore cautioned that the application of these structures should come only after sound engineering judgment has been applied. Furthermore, as an overall treatise covering a wide variety of line applications, this document cannot conceivably satisfy all conditions. The user should bear in mind that often there will be specific local conditions and requirements that may dictate design and usage conditions that differ from those described herein.

The task committee is grateful for the dedicated effort of all who participated in the development of this manual.

ACKNOWLEDGMENTS

ASCE Task Committee on Recommended Practice for Wood Pole Structures for Electrical Transmission Lines

James M. McGuire, P.E., *Chair*
Otto J. Lynch, P.E., *Vice Chair*
Vicki Schneider, P.E., *Secretary*
Yair Berenstein, P.Eng.
Nelson G. Bingel III
Ryan Bliss, P.E.
Clinton Char, P.E.
Thomas J. Christy, P.E.
Robert Erickson
Jim Fixsen, P.E.
Kristie Jackson Green, P.E.
Justin L. Johnson, P.E.
Devang L. Joshi, P.E.
Robert Kluge, P.E.
Kishor Kumar
Michael D. Miller, P.E.
Jeffrey J. Morrell
Thad A. Mumm, P.E.
Rob Nelson
David E. O'Claire, P.E.
Wesley J. Oliphant, P.E.
Jaron T. Reay, P.E., S.E.
Erik A. Ruggeri, P.E.
Katie Shaddix, P.E.
Larry Vandergriend, P.E.
Timothy P. Wachholz, P.E., S.E.
Kimberly Waggoner, P.E.
Vahid Zakeri, P.E.

The committee wishes to thank the members of the Peer Review Committee for their assistance and candid observations. It has been a pleasure to work with these individuals. Their contributions are greatly appreciated.

Gary Bowles, *Chair*
Mark Jurgemyer
Martin Rollins
David Endorf, P.E.
Stacey W. Sheffield, P.E.
Bryce Barton, Ph.D., P.E.
Terry Burly
R. Dan Seale
Steven G. Walker, P.E.

DEFINITIONS

ANCHOR GUY: Guy wire attached to a wood pole structure on one end and an anchor installed into the ground on the other end, to provide increased strength and/or stability to the transmission structure and foundation system.

ANGLE STRUCTURE: Structure where the conductor direction changes. This structure supports similar vertical and transverse loads as a TANGENT STRUCTURE, as well as significant transverse loads resulting from wire tensions being applied at an angle.

BACKFILL: Material such as soil or crushed stone that is placed to fill an excavation and provide support to the foundation system.

BUCKLING: Failure of the wood pole or other structural member to resist the applied compressive stress in its current geometric shape.

CLEARANCE: The clear distance between two objects measured surface to surface.

CONDUCTOR: A wire or cable suitable for and intended to carry electric current.

CONNECTION: The method by which loads are transferred between structural members or components.

DEAD-END STRUCTURE: Structure designed as a termination point for wires and capable of supporting the full tension loads of all wires removed on one face.

DESIGN LOADS: The loads that the structure is designed to support, comprised of NOMINAL LOADS multiplied by a specified LOAD FACTOR.

DESIGN STRENGTH: The capacity of a structural member to resist the DESIGN LOADS. It is calculated by multiplying the NOMINAL STRENGTH by the appropriate STRENGTH FACTOR.

DIRECT EMBEDDED POLE: A structure that utilizes the lower section of the pole as the primary foundation system by setting it a specified distance below groundline.

ENGINEER: An agent of the owner who is responsible for the design and is qualified by engineering education, experience, and/or professional registration to complete the design.

FRAMED STRUCTURE: Any multipole structure that is constructed using crossarms, braces, or other structural members. A simple frame structure uses only a crossarm that is pinned to each pole, whereas a braced frame structure uses knee, vee, and/or X-braces to increase the structure stiffness and capacity to support the DESIGN LOADS.

GROUNDING SYSTEM: A CONDUCTOR or system of conductors intended to provide a path for electric current to travel to earth.

GROUNDLINE BENDING MOMENT: Moment resulting from horizontal and offset loads applied at heights aboveground.

GUYED STRUCTURE: Any structure that relies on SPAN GUY wires or ANCHOR GUY wires to provide the necessary strength and stability to the structure or foundation system to support the DESIGN LOADS.

INSULATOR: Structural member composed of insulating material and designed to support a conductor physically while providing the required electrical CLEARANCE to other structural members.

JOINT USE STRUCTURE: Any structure that supports the facilities of two or more entities.

LAMINATED UTILITY STRUCTURES: Glued laminated timber structures that are an engineered, stress-rated product of a timber laminating plant and made up of assemblies of specially selected and prepared wood laminations securely bonded together with adhesives. The grain of all laminations is approximately parallel longitudinally. They may comprise pieces end-joined to form any length or pieces placed or glued edge-to-edge to make wider pieces.

LOAD FACTOR: A multiplier used with NOMINAL LOADS to introduce a level of reliability to the design and establish the DESIGN LOADS.

NOMINAL LOADS: Forces that result from meteorological conditions, physical configuration, construction activity, weight of structural members, wires, and equipment.

NOMINAL STRENGTH: The calculated capacity of a structural member to resist the effects of applied loads. This capacity is multiplied by a STRENGTH FACTOR to determine the DESIGN STRENGTH of the structural member.

OVERLOAD FACTOR: See LOAD FACTOR.

OWNER: The person, entity, or designated representative responsible for the design, fabrication, installation, purchase, and maintenance of the transmission structure.

QUALITY ASSURANCE: Administrative and procedural activities implemented in a quality system to minimize or prevent mistakes, problems, and deficiencies and fulfill quality requirements and goals for a product, service, or activity.

QUALITY CONTROL: The process by which individuals or entities review and test the quality of each aspect of a product, service, or activity to uncover defects.

RAKE: The amount of horizontal pole top displacement created by installing a pole tilted out of plumb. It is used to visually counteract pole deflection such that the pole will appear plumb under a specified load condition.

SELF-SUPPORTING STRUCTURE: Any structure that is capable of providing the necessary strength and stability to support the DESIGN LOADS without the aid of an ANCHOR GUY, SPAN GUY, or other external restraint.

SHIELD WIRE: Wire installed above the conductors for lightning protection and fault current return. Other terms commonly used include static wire, overhead ground wire, and optical ground wire (OPGW).

SPAN GUY: Guy wire attached to a structure on one end and another supporting structure on the other end, to provide increased strength and stability to the transmission structure.

STRAIN STRUCTURE: Structure similar to a DEAD-END STRUCTURE, where wire tensions are transferred directly to the structure but capable of resisting only unbalanced/differential tensions with all wires intact.

STRENGTH FACTOR: A multiplier used with the NOMINAL STRENGTH of a structural member to account for variations in the material properties, introduce a level of reliability to the design, and establish the DESIGN STRENGTH.

STRUCTURE FRAMING DRAWINGS: The drawing set prepared by the ENGINEER that contains the details and specifications necessary for the proper assembly and erection of the wood pole structure.

TANGENT STRUCTURE: Structure having minimal line angle. This structure primarily resists the vertical weight of the wires and any accumulated ice and the transverse loads caused by wind.

THROUGH-BORING PROCESS: Process that involves boring many small holes through the pole in a specific pattern at the groundline zone before application of the wood preservative. This process allows the preservatives to penetrate the heartwood of the pole and saturate the wood material in the groundline zone, providing enhanced protection against wood decay.

WIRE: All electrical conductors, shield wires, optical ground wire (OPGW), messengers, and communication cables attached to a transmission structure.

CHAPTER 1

STRUCTURAL CONFIGURATIONS AND POLE APPLICATIONS

1.1 GENERAL

Round wood poles have been used as structural members for centuries and account for most overhead utility line construction. Trees are a renewable resource, are readily available, and when harvested can be treated with preservatives that help resist decay. Utility workers can climb wood poles; the poles are nonconductive and easily modified in the field by drilling or sawing.

Utility lines can also be constructed using poles manufactured from other materials. The material chosen depends on considerations such as cost, availability, performance, and preference. When wood poles are selected as the primary structural component, it is essential that the appropriate material properties are accounted for during the design process. The species, original preservative treatment, pole length, class, and embedment depth are important aspects of design.

Consideration should also be given to the type of application and structure configuration that will be used. Designing structures based only on the pole's bending strength may not be adequate for applications requiring guyed structures or switch structures, where buckling strength and deflection should also be considered. Different species are more readily available in certain areas and have unique strengths and characteristics that may dictate the preferred material type for specific applications.

In some instances, a rule of thumb is used to install or select poles and design wood pole lines. This approach does not ensure sound engineering practices are followed. The information in this manual is intended to give the reader basic knowledge of the principles and methods for the design and use of wood poles for overhead utility line structures using sound engineering practices.

Although not part of this design document, information regarding laminated wood structures is provided in Appendix C.

1.2 CONFIGURATIONS

Wood poles are primarily used in one or more of the following structural configurations:

- Self-supporting,
- Guyed, and
- Framed.

These configurations may include transmission, distribution, and/or communication wires.

1.2.1 Self-Supporting Structures

The most common use of wood poles for overhead lines is a self-supporting structure directly embedded into the earth with soil or aggregate compacted around its base. Typical examples are shown in Figures 1-1 and 1-2. Self-supporting structures are cantilevered installations designed to withstand various combinations of vertical, transverse, and longitudinal loads as an unsupported beam column. Although shear and torsion loads create stresses on the structure, the design of this type of structure is generally controlled by the bending stresses caused by transverse and longitudinal loads. Horizontal loads are usually the result of both wire tension and wind forces on the structure, equipment, and wires. Offset vertical loads on one side of a pole that are not balanced by a similar load on the other side also contribute to bending stresses. Another example of offset vertical loads is created when a structure leans or is deflected, so vertical loads caused by equipment, conductors, and ice are no longer balanced. Self-supporting wood pole structures are typically tangents or are only capable of supporting small line angles.

1.2.2 Guyed Structures

Another common configuration of wood poles for overhead lines is a guyed structure. To reduce the bending stresses and deflection associated with self-supporting structures, guy wires can be installed to transmit the horizontal loads imposed on the structure to other supporting structures or anchors installed in the ground. Although guy wires significantly reduce

Figure 1-1 (left). Self-supporting post insulator tangent structure supporting multiple circuits.
Source: Jaron Reay, reprinted with permission.

Figure 1-2 (right). Self-supporting wishbone tangent structure supporting multiple circuits.
Source: Jaron Reay, reprinted with permission.

bending stresses in the wood pole structure, the vertical component of the tension in the anchor guy induces additional vertical loading on the pole. In addition to this document, ASCE Manual of Practice No. 91, *Design of Guyed Electrical Transmission Structures* (ASCE 1997), provides more information on this topic.

Figures 1-3 through 1-9 depict various guyed transmission structures. A guyed structure is typically analyzed as a system to determine the overall structural capacity. The guy wire size, orientation, pretension, and load capacity should be specified along with the pole size and attachments on the pole.

Figure 1-3 (left). Single pole anchor guyed angle structure.
Source: Otto Lynch, reprinted with permission.

Figure 1-4 (right). Single pole anchor guyed dead-end structure.
Source: Robert Erickson, reprinted with permission.

Figure 1-5. Single pole span guyed structure.
Source: Otto Lynch, reprinted with permission.

Figure 1-6 (left). Three-pole guyed dead-end structure.
Source: Robert Erickson, reprinted with permission.

Figure 1-7 (right). Three-pole guyed dead-end switch structure.
Source: Jaron Reay, reprinted with permission.

Figure 1-8 (top) and Figure 1-9 (bottom). Two views of three-pole guyed angle structures.
Source: Jaron Reay, reprinted with permission.

1.2.3 Framed Structures

Framed structures consist of more than one pole that can be connected in a variety of configurations. The H-frame is the simplest version. In this structure two adjacent poles are connected to the same crossarm as shown in Figure 1-10. Since the crossarm connection is usually pinned, load sharing occurs between the two poles. However, the distribution of stress along the length of each pole is similar to self-supporting single pole structures.

Framed structures with added X-braces or knee braces act as a rigid, moment-resisting system. Horizontal loads are transmitted to the ground through the total stiffness of the structural support system. This stiffness is achieved by adding bracing members with pinned connections, moment-carrying connections, or a combination of the two. Examples of braced H-frame structures are shown in Figures 1-11 and 1-12.

It should be noted that the structure is cantilevered in the longitudinal direction, perpendicular to the plane of bracing, and must be designed to support the bending moment created from longitudinal loading.

Figure 1-10. H-frame structure.
Source: Jaron Reay, reprinted with permission.

Figure 1-11 (left). Braced tangent H-frame structure.

Figure 1-12 (right). Braced strain H-frame structure.
Source: Otto Lynch, reprinted with permission.

Similar to guyed structures, framed structures are typically analyzed as a system to determine the overall structural capacity. The engineer is therefore responsible for determining the size, orientation, and connection details of all members in the frame.

1.3 ELECTRIC UTILITY APPLICATIONS AND STRUCTURAL CONFIGURATIONS

The structural configurations mentioned previously lend themselves to several common wood pole structure applications.

Self-supporting structures are commonly used in tangent construction where the wire tension creates relatively little or no transverse or longitudinal loading on the pole. Governing loads on self-supporting structures are transverse loads owing to wind on bare or iced wires. Because they are unsupported cantilever structures, the overall load-resisting capacity is somewhat small compared to other structure types. As the total bending stress typically controls the design of self-supporting structures, the wood pole class, length, embedment depth, and load configuration are all critical components for the engineer to consider when determining the design strength of the structure.

Guyed structures are commonly used for installations with line angles and/or unequal tensions that use smaller pole sizes than would be required without guy wires. In most cases, guying is required because the loading exceeds the cantilever capacity of the unguyed poles. The structures may be guyed at some or all wire-attachment positions, depending on the loading. When designing guyed structures, the engineer should consider constraints that include but are not limited to right-of-way, easement, existing infrastructure, archeological and environmental conditions, and soil type.

Braced H-frame structures typically support larger transverse and vertical loads than single-pole self-supporting structures, resulting in longer allowable span lengths and/or the use of larger wire sizes. These structures can also be designed to carry significant longitudinal loads by guying or other methods. Because structures that are framed and braced create a different type of loading action at the groundline, engineers must consider uplift and the bearing capacity of the soil as part of the structure design.

Combinations of self-supporting, guyed, and framed structures can be used to optimize the benefits of each type. Guyed H-frames provide increased longitudinal capacity as an anti-cascading structure in long cross-country lines or where differing span lengths create longitudinal load imbalances. Large-angle dead-end structures guyed only in one direction are commonly used at locations where the wires on one side are full tension and the wires on the other side are reduced tension.

A variety of terms are used to describe the different types of conductors and wires attached to a pole or the function of the attachment. Common terms include transmission, distribution, electric, supply, underbuild, communication, fiber, phone, cable TV, and joint use. For the purpose of this text on recommended practice for the design and use of wood pole structures for electrical transmission lines, all wire types are included and not just those transmitting power.

Typically, transmission lines carry voltage at or above 69 kilovolts (kV); distribution systems typically operate at less than 69 kV. Transmission systems feed substations where the voltage is stepped down to the distribution level. Distribution wires are routed to the homes or businesses of individual consumers where the voltage is stepped down to a lower voltage useful for powering their service. Various pole construction configurations are used for each type of attachment or combination of attachments, thus the various terms. Please note, however, that these are common terms used in the industry, not definitions. Figures 1-13 through 1-15 present examples of structures that support wires energized at distribution-level voltages.

The term *underbuild* is commonly used to refer to distribution circuits built under transmission circuits on the same structure, as shown in Figure 1-16. The transmission and distribution circuits have different construction requirements because of their operating voltages.

Figure 1-13 (left). Distribution riser structure.

Figure 1-14 (middle). Distribution angle structure.

Figure 1-15 (right). Distribution tangent structures.
Source: Robert Erickson, reprinted with permission.

Figure 1-16. Single-circuit transmission line with underbuild.
Source: Robert Erickson, reprinted with permission.

A double-circuit line refers to two separate transmission circuits built on the same structure, either with or without underbuild, as shown in Figures 1-17 and 1-18. The two circuits can be on opposite sides of the structure or positioned one above the other. A line with one transmission circuit and one distribution circuit, as well as communication wires or other wire attachments, is typically not considered double circuit but rather a single circuit with underbuild.

It is common to have various combinations of transmission, underbuild, and communication on the same structure, as shown in Figures 1-19 and 1-20.

Figure 1-17. Double-circuit transmission line structure with underbuild. Source: Jaron Reay, reprinted with permission.

Figure 1-18. Double-circuit transmission line structure without underbuild. Source: Robert Erickson, reprinted with permission.

Figure 1-19 (left) and Figure 1-20 (right). Two views of structures with transmission, distribution underbuild, communications, and attachments.
Source: Jaron Reay, reprinted with permission.

REFERENCE

ASCE. 1997. *Design of guyed electrical transmission structures*, MOP 91. Reston, VA: ASCE.

CHAPTER 2
INITIAL CONSIDERATIONS

2.1 GENERAL

Several considerations need to be addressed in the design of wood pole structures to develop acceptable, efficient solutions. Each variable must be adequately defined either by calculation, testing, or assumption (based on history and engineering judgment) so that the design can be developed with an acceptable level of confidence.

The following is a brief description of factors to consider when determining if wood pole structures are the best solution for a given application.

- Initial Installed Cost—This is the cost to design, procure, and install the structure or construct the line.
- Availability—The availability of various wood pole lengths, classes, and species will influence if wood pole structures provide an effective and efficient line design.
- Structural Resiliency—The higher mean strength value of wood compared to other materials suggests that a line built with wood structures could be more resilient in withstanding storm loading. Refer to Appendix A for a more complete discussion on resiliency.
- Life-Cycle Costs—There are costs associated with maintenance and replacement of the structure for a specified service life. Wood can have lower life-cycle costs than other materials in many areas of the United States, especially when utility companies have effective pole maintenance programs.
- Unique Applications—The engineer must determine whether the wood pole is able to support unique design applications—for example, crossings or long spans that require unusually tall structures,

Figure 2-1. Woodpecker entrance hole.
Source: Robert Erickson, reprinted with permission.

switch structures requiring minimal deflection, and locations where only a minimal structure footprint is allowed.

- Insect/Animal Vulnerability—The structure must maintain its structural capacity in areas where animal and insect degradation is common, with maintenance options to address termite damage, bear clawings, bird nests, and woodpecker borings. A woodpecker entrance hole is shown in Figure 2-1.
- Pole Degradation Caused by the Environment—The desired service life of the structure is based on the selection of the species and preservative treatment for a specific geographic location.
- Environmental Exposure—Some poles may be installed in extreme environments, such as in swamps and along riverbanks. The engineer should determine the optimum species, original preservative treatment, and foundation options in these cases.
- Environmental Impact—These considerations range from the energy required to manufacture the structure to the possible migration of chemical treatments into the environment during installation and while in service and the ability to recycle the pole at the end of its service life.
- Fire Resistance—The ability of the structure to withstand brush fires and options to protect wood poles from fire damage.
- Product Uniformity—The design should be able to accept the normal variation in dimensions of wood poles.

- Transportation and Delivery Lead Time—The weight of the pole, the ability to transport the required pole length to the project site, and the time required for a provider to deliver the structure after receipt of an order are all considerations.
- Weight and Crane Requirements—The contractor will need the right equipment to install the pole, considering the overall length and weight of the structure.

2.2 PHYSICAL CHARACTERISTICS

Before proceeding with design work and ultimately procuring wood poles for construction, the engineer must consider several physical characteristics. Wood species is of primary importance. Different wood species have different fiber strengths, perform differently in various environments, have different weights for a given length and class, and are more readily available in certain regions than in others. All these considerations will affect the design, constructability, cost, and/or maintenance of the transmission line, so the wood species must be established as an initial consideration based on specific project and owner requirements.

The type of preservative used and the range of its application are also important. Different environments may dictate which type of preservative treatment will be most effective. The engineer must also establish whether the treatment is applied to a zone on the pole or if the entire pole should be treated. Typically, the entire pole receives preservative treatment; some species may receive additional preparation to enhance preservative retention in the groundline zone where decay most commonly occurs.

The embedment depth is another important initial characteristic. Embedment depths should be calculated based on local soil conditions, performance and reliability requirements, and expected pole loadings. The embedment depth may also be impacted by the structure configuration being used. A framed structure installed where side slope exists may require that the uphill pole be embedded deeper than required by engineering calculations. Establishing the embedment depth can impact other physical characteristics such as the preservative treatment zone and the design strength of the pole.

The engineer must also establish if the wood poles will be procured with predrilled holes to accommodate mounting equipment and other attachments. Greater accuracy in structure framing, improved constructability, and less costly construction labor are typical when the pole manufacturer provides wood poles with predrilled holes. Information about the location, size, and orientation of holes to be drilled must be clearly communicated to the manufacturer at the time of procurement.

Each of the physical characteristics mentioned here will be discussed in greater detail in subsequent chapters of this manual.

2.3 CLEARANCES AND POLE LENGTH

The length of wood poles used to support overhead lines is established based on the required clearances to the attached facilities and the required embedment depth. Required clearances along the span from wires, equipment, and other energized parts to ground or other objects and installations are specified by local, state, and federal codes such as California's General Order 95 (GO 95) (California PUC 2018) and the National Electrical Safety Code (NESC) (IEEE 2017). Configurations are also determined based on the required spacing between facilities attached to the wood pole structure. The engineer must therefore develop the structure configurations and an overhead line design that conforms to all clearance and spacing requirements and combine those designs with the required embedment depth of each structure to define the required length of each wood pole.

2.4 LOADS AND POLE CLASS

The magnitude and direction of all the design loads must be considered when determining the required class of a wood pole. These loads should comply with the requirements of the NESC (IEEE 2017), ASCE Manual of Practice No. 74 (ASCE 2010), state and local codes, and other applicable codes and design specifications. Any other design load cases required by

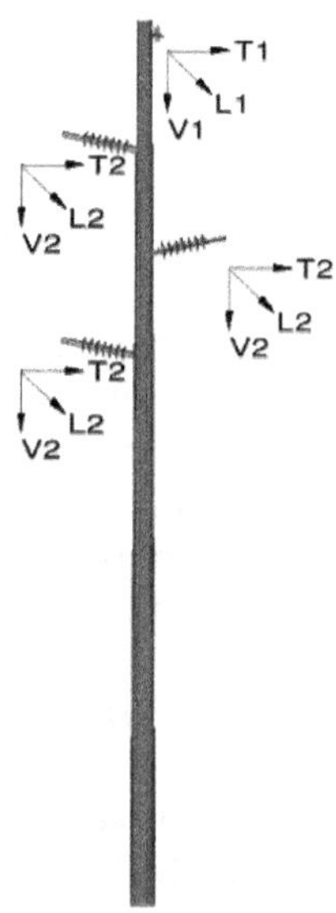

Figure 2-2. Typical structure loading diagram.
Source: Otto Lynch, reprinted with permission.

the owner, as well as the deadweight loads of any attached equipment, should also be incorporated. When specifying loads applied to a wood pole structure, it is helpful to use a load tree and/or load table in longitudinal/vertical/transverse (LVT) format as shown in Figures 2-2 to 2-4.

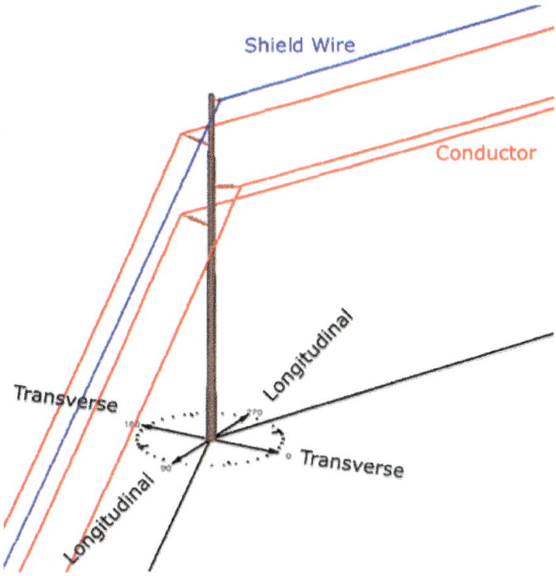

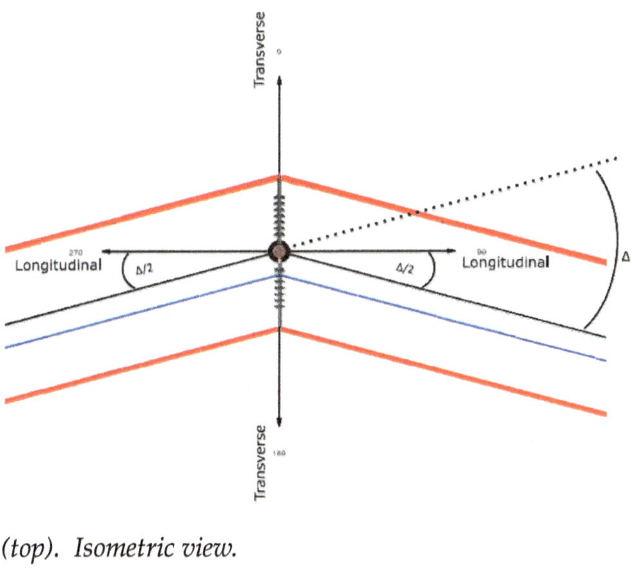

Figure 2-3 (top). Isometric view.

Figure 2-4 (bottom). Plan view.
Source: Otto Lynch, reprinted with permission.

Once all applicable load cases have been considered, the wood pole class can be selected to provide the design strength to adequately support the design loads.

2.5 DEFLECTIONS

Deflection limitations may be required to maintain proper clearances from the structure and its attachments to other objects and/or the edge of rights-of-way or easements. They may also be necessary because of equipment operation tolerances, such as those for switches. These deflection limitations and the conditions under which they should be met should be considered when designing a wood pole structure. Figures 2-5 and 2-6 are examples of structure deflection tests.

Figure 2-5. H-frame load and deflection test.
Source: Hughes Brothers, reprinted with permission.

Figure 2-6. Single pole load and deflection test.
Source: Osmose Utilities Services, Inc., reprinted with permission.

2.6 TRANSPORTATION AND ERECTION

Consideration should be given to how the structure sites will be accessed and how the poles will be delivered and erected. Restrictions on access may limit the length of poles that can be delivered to a site or limit the weight of a pole that can be erected based on the size of the construction equipment needed.

2.7 ATTACHMENTS

Attachments to wood pole structures can include but are not limited to shield wires, insulators, structural supports for conductors, guy wires, ground wires, down lead clips, supports for equipment such as switches or distribution equipment, woodpecker wrap, danger signs, aerial markers, and similar equipment used by joint use entities. Figures 2-7 and 2-8 show examples of attachments.

The decision on where to locate the attachments is driven by the overall functionality of the pole and the need to maintain all clearance requirements. The engineer should specify the hole locations, diameters, and patterns required to attach the insulators and other hardware and equipment to the structure. Chapter 5 describes the typical connection details for different types of component attachments.

Figure 2-7 (top). Porcelain horizontal post with a clamptop.

Figure 2-8 (bottom). Shield wire bolted dead-end.
Source: Robert Erickson, reprinted with permission.

2.8 GUYING

The engineer should evaluate all pertinent information regarding the guying of the structure. This includes guy wire size, modulus of elasticity, rated breaking strength, stranding, maximum allowable load, guy quantity restrictions for anchors, guy angle limits for attachment hardware, and restrictions imposed by land-use agreements and terrain conditions. In addition to this document, refer to *Design of Guyed Electrical Transmission Structures* (ASCE 1997). An example of connections for guying is shown in Figure 2-9.

Figure 2-9. Connections for guying.
Source: Robert Erickson, reprinted with permission.

The engineer should confirm the structural capacity of all guys, attachment hardware, and anchors. Multiple guy wire configurations, sizes, and components should be considered to optimize the design of the pole.

2.9 CLIMBING AND MAINTENANCE

When designing a wood pole structure and specifying a pole size and preservative treatment, the engineer must consider the ability to access those areas of the pole where hardware is attached. Poles are accessed by bucket truck, climbing, or at times by a ladder.

If climbing is considered, Occupational Safety and Health Administration (OSHA) regulations and appropriate owner safety practices should be addressed. Providing adequate clearance for line personnel from energized parts needs to be considered in the design whether work will be accomplished while climbing the pole or working from a bucket truck, as shown in Figure 2-10. Water-borne original preservative treatments may make climbing difficult, so the type of treatment must also be considered. Chapter 9 has additional information regarding wood pole inspection and maintenance.

Figure 2-10. Climbing space on pole.
Source: Robert Erickson, reprinted with permission.

2.10 GROUNDING

Wood pole structures should be effectively grounded to provide protection from pole fires and lightning strikes and electrical fault protection for the power line and substation facilities. Many types of grounding systems can be employed depending on the resistivity of the soil, the electrical requirements of the transmission line, and owner specifications. Two common grounding systems employed are driven ground rods and wrapping the ground conductor around the embedded portion of the pole in a specific pattern. Figures 2-11 and 2-12 show pole grounding without and with rods, respectively. Grounding requirements should be clearly specified by the owner and incorporated into the design. See IEEE P751/D2, *Draft Guide for Wood Structures Used for Overhead Electric Transmission Lines* (IEEE 2018) for more information on grounding.

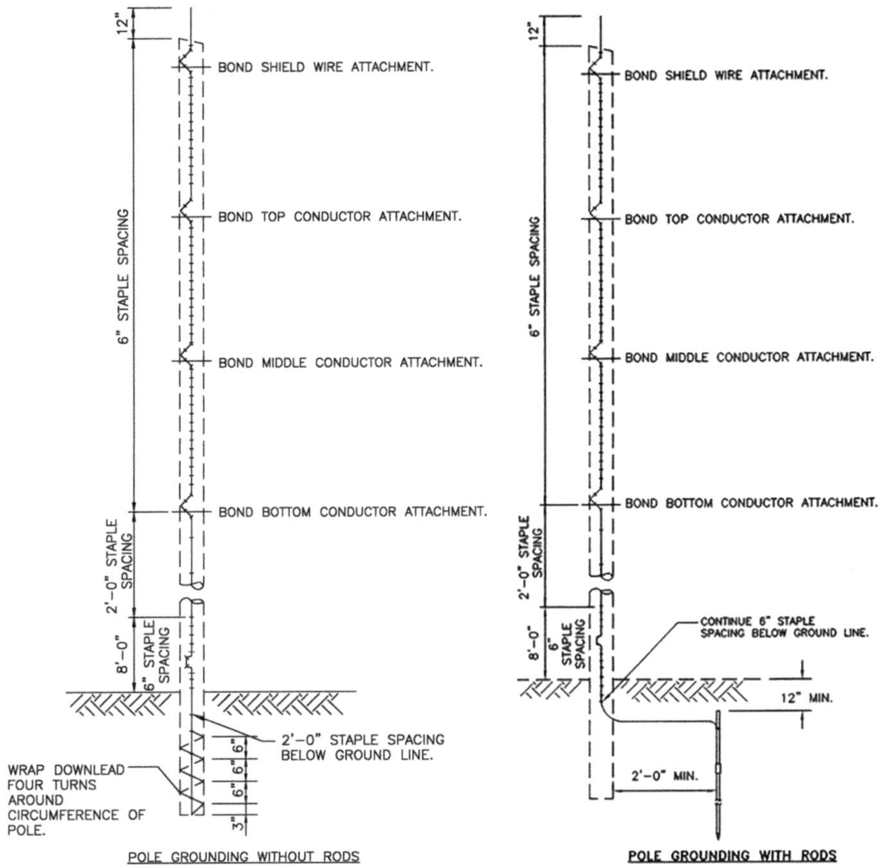

Figure 2-11 (left). Pole grounding without rods.

Figure 2-12 (right). Pole grounding with rods.
Source: Great River Energy, reprinted with permission.

REFERENCES

ASCE. 1997. *Design of guyed electrical transmission structures*, MOP 91. Reston, VA: ASCE.

ASCE. 2010. *Guidelines for electrical transmission line structural loading*, MOP 74. Reston, VA: ASCE.

California PUC (Public Utilities Commission). 2018. *Rules for overhead line construction*, General Order No. 95. Accessed November 18, 2018. http://www.cpuc.ca.gov/gos/GO95/go_95_startup_page.html.

IEEE (Institute of Electrical and Electronic Engineers). 2017. ANSI Standard C2, *National electrical safety code*. New York: IEEE.

IEEE. 2018. 751, *Draft guide for wood structures used for overhead electric transmission lines*. New York: IEEE.

CHAPTER 3
MATERIALS

3.1 GENERAL

This chapter offers information on the mechanical properties, applicable standards, and specifications used to manufacture wood poles. Preservative treatments and conditioning of the pole are discussed briefly and covered in greater detail in Chapter 7.

3.2 PHYSICAL PROPERTIES

The physical properties for wood poles in the utility industry in North America are found in the latest versions of American National Standards Institute (ANSI) Standard O5.1 (ANSI/AWPA 2017) and Canadian Standards Association (CSA) Standard CAN/CSA-O15 (CSA 2015). Laminated wood pole properties are based on ANSI Standard O5.2 (ANSI/AWPA 2012) and ANSI Standard 117-2015 (ANSI/APA 2015).

3.2.1 Designated Fiber Strength

Substantial work to determine and specify general wood properties began during the early 1900s, but the first major studies of wood pole properties were done in the early 1960s. *Strength and Related Properties of Wood Poles* (Wood et al. 1960) presented the results of a large ASTM study on various tests of small wood pole samples and full-size utility poles (25 to 35 ft) made of a variety of species. These data were then used to establish designated wood fiber strengths by the Forest Products Laboratory in FPL 39—*Derivation of Fiber Stresses from Strength Values of Wood Poles* (Wood and

Markwardt 1965). These values were adopted in 1963 in ANSI O5.1. Since that time a number of additional full-scale pole tests, including a large sample of transmission-size poles, were conducted by the Electric Power Research Institute in the 1980s (EPRI 1981, 1985, 1986a, 1986b). A summary of the test results is published in Annex C of ANSI O5.1 (ANSI/AWPA 2017).

The Accredited Standards Committee O5 (ASC O5) is responsible for the series of standards that address wood poles and crossarms. In the late 1990s, the committee combined the full-scale test data from both projects and eliminated the small, clear specimen data. The objective was to determine if any of the 1963 designated fiber strength values should be changed because of the additional EPRI testing.

A new set of adjustments to the test data was applied to align data from different test projects to a similar basis. This derivation was documented in "Evaluation of Fiber Stress Values for Wood Utility Poles in ANSI Standard O5.1" (Rollins 2004) and showed that no change to the existing fiber strengths was warranted.

Table 1 in ANSI O5.1-2017 shows the designated fiber strengths for all species of wood poles. The effects on the pole strength from conditioning for preservative treatment have been accounted for in the designated fiber-strength values.

It is important to note that the designated fiber strengths for Southern pine, Douglas fir, and Western red cedar are mean values with a coefficient of variation equal to 20% as explained in footnote 4 of ANSI O5.1 Table 1.

Tables 3 through 10 of ANSI O5.1 define the circumference requirements for each pole class and length of the most common wood pole species by specifying the minimum circumference requirements at 6 ft from the butt and at the tip. The values are correlated with the fiber strength of each species so that all poles of the same length and class will have similar groundline moment capacity.

When wood transmission structures are designed, poles are assumed to have the minimum required dimensions for the specified class. A field study collected data on the actual dimensions of more than 22,000 poles during manufacturing and was first published in the 2015 edition of ANSI O5.1. This data validated that assuming minimum dimensions is conservative and identified the extent of oversize that exists in the average pole.

It is important to understand that designated fiber strengths specified in ANSI O5.1 are groundline values (Wolfe and Kluge 2005). Annex A of ANSI O5.1 also explains that the fiber strength of round utility wood poles decreases with height aboveground (Bohannan 1971). This is particularly important for guyed or braced structures where the maximum stress may occur well above the groundline.

However, the oversize of the average pole compared to minimum dimension requirements was found to offset the fiber-strength reduction above the groundline for poles that are 55 ft and shorter. Therefore, the fiber strength height effect does not apply to these poles. For single poles that are 60 ft and longer, where the maximum stress point is above the groundline, the fiber strength height effect should be considered. Equations in Annex A of ANSI O5.1 specify the necessary reduction in fiber strength. Section A.3 of Annex A of the standard explains how full-scale testing has shown that multipole structures of any length may not need to reduce capacity because of the fiber strength height effect.

3.2.2 Modulus of Elasticity

The response of structures to applied loads is a function of the strength and stiffness properties of the materials comprising the structures (CEA 1994). For wood poles, the most important measure of stiffness used in analysis and design calculations is the modulus of elasticity (MOE). These stiffness values for wood poles have been published in Table 1 of ANSI O5.1 since the 2017 edition. Proper application is critical because MOE values, together with reliable structure and component geometry information, are the basis for evaluating important issues such as

- Load sharing/stress distribution among components of a structure (Figure 3-1);
- Structure deflection, including nonlinear large-deflection effects (aka P-delta effects);
- Column buckling capacity;

Figure 3-1. Testing of framed structure under transverse loading.
Source: Hughes Brothers, reprinted with permission.

- Right-of-way (ROW) width and lateral clearance adequacy including consideration of structure deflection, insulator swing, and conductor blowout; and
- Ability to absorb unbalanced dynamic loads.

3.3 PRESERVATIVE TREATMENT

All materials degrade when the proper conditions exist, and wood poles are no exception. The wood species used to support electric lines differ in their typical decay patterns. The service life of poles can be enhanced by an original preservative treatment, usually incorporating pressure to force the chemical deeper into the wood. The standards for conditioning and treatment of poles in the United States are promulgated by the American Wood Protection Association (AWPA). The AWPA specifications are based on the decay risk to which the wood will be exposed. Utility poles are in Use Category 4, and there are three subcategories (A–C) with increasing degrees of decay potential. The AWPA produces a decay risk map to help guide utilities in choosing the proper Use subcategory for their specific service conditions. The Canadian Standards Association takes a similar approach, using decay risk to choose an appropriate treatment level.

The AWPA and CSA standardize preservative treatments. There are two basic categories of treatments: oil-soluble or miscible systems, and water-soluble systems. Preservative systems currently specified for use on utility poles include chromated copper arsenate (CCA), pentachlorophenol (PCP), creosote (CR), alkaline copper quat (ACQ), ammoniacal copper zinc arsenate (ACZA), copper azole (CA), and copper naphthenate (CuN). These systems are considered biologically equivalent at the levels specified in the standards, although they may differ in other performance attributes. At present, a majority of poles are treated using either PCP, CR, or CCA.

In addition to chemical compositions, the treatment standards place limits on heating and pressure to minimize risk of damage during treatment. They also include specific pretreatment processes such as conditioning, pre-boring for all attachments and making any cuts prior to preservative treatment. Additional processes such as incising and through-boring are included for species that are more resistant to absorbing preservative treatment. Refer to ANSI O5.1 Table 1, footnote 5 for 5% reduction of designated fiber strength for through-bored treatment for Douglas-fir poles.

Treatment quality is based on the penetration depth and the amount of preservative retained in targeted sections of the pole. Penetration and retention are determined by removing increment cores from the wood poles after treatment and performing a chemical analysis. The AWPA standards are considered minimums to achieve good performance. Utility specifications can exceed these criteria, but they should be certain that the more stringent specification actually provides a performance gain.

REFERENCES

AITC (American Institute of Timber Construction). 2010. AITC 117, *Standard specifications for structural glued laminated timber of softwood species.* Centennial, CO: AITC.

ANSI/APA (The Engineered Wood Association). 2015. ANSI 117-2015, *Standard specification for structured glued laminated timber of softwood species.* Tacoma, WA: APA.

ANSI/AWPA (American Wood Protection Association). 2012. ANSI-O5.2, *Structural glued laminated timber for utility structures.* Birmingham, AL: AWPA.

ANSI/AWPA. 2017. ANSI-O5.1, *Wood poles—specifications and dimensions.* Birmingham, AL: American Wood Protection Association.

Bohannan, B. 1971. *Height-strength relationship for round timbers.* Working document for ANSI-O5.1 Committee.

Canadian Standards Association (CSA). 2015. Standard CAN/CSA-O15, *Wood utility poles and reinforcing stubs.* Mississauga, ON: CSA Group.

CEA (Canadian Electricity Association). 1994. Standard 119 D 1010, *Strength and stiffness of new wood utility poles.* Ottawa, ON: CEA.

EPRI (Electric Power Research Institute). 1981. EL-2040, *Probability-based design of wood transmission structures.* Palo Alto, CA: EPRI.

EPRI. 1985. EL-4109, *Wood pole properties, Vol. 1, Southern pine data.* Palo Alto, CA: EPRI.

EPRI. 1986a. EL-4109, *Wood pole properties, Vol. 2, Douglas-fir data.* Palo Alto, CA: EPRI.

EPRI. 1986b. EL-4109, *Wood pole properties, Vol. 3, Western red cedar data.* Palo Alto, CA: EPRI.

Rollins, M. 2004. "Evaluation of fiber stress values for wood utility poles in ANSI Standard 05.1." *Report to the Standard 05.1 Committee.* Vancouver, WA: North American Wood Pole Council.

Wolfe, R. W., and R. O. Kluge. 2005. "Designated fiber stress for wood poles." *Gen. Tech. Rep. FPL 158.* Madison, WI: USDA Forest Service, Forest Products Laboratory.

Wood, L. W., E. C. O. Erickson, and A. W. Dohr. 1960. *Strength and related properties of wood poles.* Philadelphia, PA: ASTM.

Wood, L. W., and L. J. Markwardt. 1965. "Derivation of fiber stresses from strength values of wood poles," *FPL 39.* Madison, WI: USDA Forest Service, Forest Products Laboratory.

CHAPTER 4
DESIGN OF WOOD POLE STRUCTURES

4.1 GENERAL

Wood poles are a cost-effective structural material that can be advantageous in the design of overhead lines. Natural wood poles have a higher degree of variability than engineered materials in terms of strength factors. The high coefficient of variability of wood poles means a large percent of the population has a greater nominal strength than used in initial design. This is discussed in more detail in Appendix A.

The intent of this chapter is to use the best material property data available from multiple sources and apply current structural engineering principles to recommend appropriate design values and considerations. The desired outcomes are accurate and reliable wood pole designs that are consistent in performance with other types of structure materials, such as concrete, tubular steel, and lattice steel when used in the design of overhead lines.

The design principles and methodologies outlined focus on the structural engineering principles of Load and Resistance Factor Design (LRFD). This format is similar to that used in most current structural engineering design methodologies including the ASCE Steel Pole Standard 48-11 (ASCE/SEI 2012) and ASCE Manual of Practice No. 123 on recommended practice for design of concrete transmission poles (ASCE 2012). Given the requirement to also incorporate legislated strengths and loadings outlined by the National Electrical Safety Code (NESC) (IEEE 2017) or General Order (GO 95) issued by the state of California (California PUC 2018), the requirements outlined in those codes must also be considered and used as prescribed in those documents. Specifications for the quality, minimum dimensions, and mean nominal strength of wood poles used for overhead

Table 4-1. Horizontal tip load.

Pole (Class)	Horizontal Load* (Pounds)	Pole (Class)	Horizontal Load* (Pounds)
H6	11,400	1	4,500
H5	10,000	2	3,700
H4	8,700	3	3,000
H3	7,500	4	2,400
H2	6,400	5	1,900
H1	5,400	6	1,500
		7	1,200
		9	740
		10	370

*Horizontal load is applied 2 ft from tip.

lines are published under the jurisdiction of the Accredited Standards Committee O5 (ASC O5), ANSI Standard O5.1 (ANSI/AWPA 2017).

4.2 STRUCTURAL DESIGN OBJECTIVE AND METHODOLOGY

Wood poles of each species are dimensionally "classed" to provide approximately equivalent mean nominal groundline moment capacity at an approximated groundline location. The average horizontal tip loads that a pole must support for each class designated in Annex B of ANSI O5.1 (2017) are provided here in Table 4-1.

Since wood poles are designated as a particular length and class, the objective of the structural design methodology outlined in this chapter is for the engineer to identify the wood pole class required for a given length, species, and framing that will have greater design strength than the design load and load effects expected to be experienced by the structure over its design life. To compare strengths of various pole materials, it is desirable to use an approximate 5% lower exclusion limit value to define "strength" with all pole materials.

The method of designing a structure with greater design strength than the maximum load and load effects is expressed in an LRFD format as

$$\emptyset R \geq \gamma Q \tag{4-1}$$

where

$\emptyset$ = Strength factor,

R = Measure of resistance (nominal strength),

γ = Load factor, and
Q = Calculated load or load effects.

For wood poles, nominal strength is the designated fiber strength, or modulus of rupture (MOR) in bending. Designated fiber strength for common wood pole species is published in Table 1 of ANSI O5.1. These values of designated fiber strengths include the effects of conditioning that occur before treatment.

Since the designated ANSI O5.1 Table 1 fiber strength values are "mean" strength values (50% lower exclusion limit, or LEL_{50}) and a 5% LEL strength value is desired for design, these designated fiber strengths need adjustment per the following equation (based on a "normal" probability density function (PDF) of the measurements of MOR):

$$MOR_{5\%} = MOR_{50\%} \, (1 - 1.645 \times COV) \tag{4-2}$$

where
$MOR_{5\%}$ = 5% lower exclusion limit,
$MOR_{50\%}$ = 50% lower exclusion limit, and
COV = Coefficient of variation as designated in Table 1, footnote 4 of ANSI O5.1.

The information provided on reliability-based design in Annex C of ANSI O5.1 (ANSI 2017) and ASCE Manual of Practice No. 111 (ASCE/SEI 2006) is limited to the use of unguyed single-pole structures. Because of the limited application of these documents, they are not widely used.

The modulus of rupture in bending reflects the maximum load-carrying capacity of a member in bending and is proportional to ultimate moment that can be supported by the pole species. Modulus of rupture is an accepted criterion of strength, although it is not a true strength because the formula by which it is computed is valid only to the elastic limit of the wood pole.

When the theoretical point of maximum stress for single poles 60 ft and longer is above groundline, Annex A of ANSI O5.1 (ANSI 2017) provides equations for adjustment of the MOR at locations along the pole, depending on the overall length. That formula for Southern pine, Douglas fir, and Western red cedar is

$$F_2 = F_1 \, (1 - 0.3 \, H/L) \tag{4-3}$$

where
F_2 = Calculated fiber strength value (MOR) at a distance H above groundline,
F_1 = ANSI O5.1 Table 1 fiber strength value ($MOR_{50\%}$),
H = Distance from groundline to a point aboveground where the fiber strength is F_2 (maximum value of H can be L/2), and
L = Length from groundline to the top of the pole.

For other species that were not included in the size study, ANSI O5.1 modifies the previous formula to

$$F_2 = F_1 (1 - 0.5\ H/L) \tag{4-4}$$

4.3 ANALYSIS AND DESIGN

As stated previously, the primary consideration is to ensure that the applied loads (with appropriate load factors) do not exceed the pole design strength. Probabilistic loads and load factors are published in *Guidelines for Electrical Transmission Line Structural Loading* (ASCE 2010). Legislated loads (such as in the NESC and GO 95) should use the load and strength factors dictated by those standards.

4.3.1 Statically Determinate Structures

For statically determinate structures, such as self-supporting and framed structures without bracing, the transverse bending moment most often controls the structure design. Axial forces are generally minimal compared to bending moment but are important contributors to secondary bending moments when the structure deflects because of applied loads. Often referred to as P-delta moments, they should not be neglected. Stresses resulting from shear loads will generally be minimal compared to bending moment and the additive moment effects of axial loads. Reactions are relatively simple to calculate for statically determinate structures.

Primary bending moment is calculated as

$$M = Q_1 D_1 + Q_2 D_2 + \cdots Q_n D_n \tag{4-5}$$

where

M = Primary bending moment,

Q = Relative distance of each applied horizontal load (or vertical loads with eccentric offsets), and

D = Relative distance of each applied horizontal load (or eccentric offset) to the cross-section being analyzed.

Figure 4-1 depicts the deflected shape of a typical tangent structure and vertical loads relative to their displacements that create the secondary moments. The best way to calculate secondary moments is by using iterative nonlinear analysis techniques provided within most pole analysis software. However, if performing calculations by hand or using a spreadsheet, an approximation of the secondary bending moments can be determined by calculating a bending moment amplification factor that is applied to the primary bending moment. The most common approximation method for

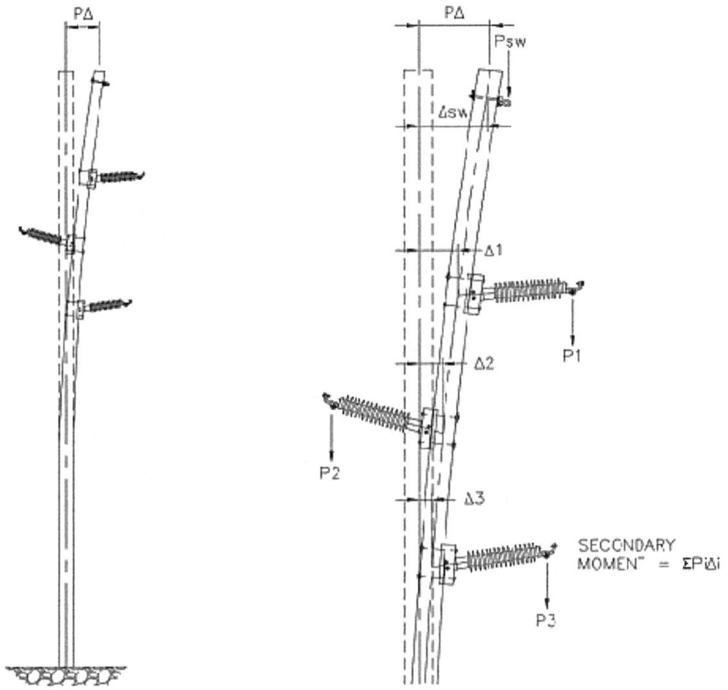

Figure 4-1. P-delta diagram.
Source: Great River Energy, reprinted with permission.

calculating an amplification factor is derived from the Gere-Carter method for determining critical buckling loads for tapered columns. The amplification factor, commonly referred to as a P-delta factor, is represented by the following equation:

$$\text{P-delta factor} = 1/(1 - (VL / P_{cr})) \tag{4-6}$$

where VL is the total vertical load on the pole (including the weight of the pole aboveground). A conservative approach sometimes used is to include the weight associated with the belowground length in the total pole weight.

P_{cr} is the critical buckling load calculated as

$$P_{cr} = P'_{cr} \times P^* \tag{4-7}$$

where

$$P^* = (D_{groundline}/D_{top})^{2.7} \tag{4-8}$$

For a self-supporting structure:

$$P'_{cr} = (\pi^2 \times MOE \times I_{top}) / (4L^2 \times 144) \tag{4-9}$$

where

MOE = Modulus of elasticity of the wood pole (psi) in Table 1 of ANSI 05.1 (2017 or later),

I_{top} = Moment of inertia at the top of the unsupported length of the column (in.^{4}),

L = Unsupported length (ft),

$D_{groundline}$ = Diameter of the pole at the groundline, and

D_{top} = Diameter of the pole at the top of the unsupported length of the column.

4.3.2 Statically Indeterminate Structures

Most guyed single poles and multipole structures are statically indeterminate, and their analysis is best accomplished by nonlinear computer analysis software. Prudent structural engineering judgment must always be used when considering a number of critical assumptions that are required with indeterminate type analyses. These include

- Fixity of connections (X-braces, crossarms, etc.),
- Initial installed guy tensions, and
- Unbraced length factors (X-braces, etc.).

With statically indeterminate structures, the resulting calculated utilization at various sections along the pole can still be determined using the calculated reactions and the same stress equations that are used for statically determinate poles.

Modifiers for the unbraced length of a compression member are known as K factors and are dependent on the fixity of end-condition assumptions made for that member. The end-condition can be different for each of the primary axes. An X-brace can have a pinned end-condition for an in-plane connection fixity and at the same time have a fixed end-condition for an out-of-plane connection fixity. Prudent engineering judgment should always be taken when selecting the appropriate K factor.

To determine the unbraced length factor for compression members of X-brace systems bending within the plane of bracing, the tension/compression members need to be connected at the crossover point, with both ends of each brace pinned. For structures constructed with parallel poles, such as a typical H-frame, the unbraced length factor would be K = 0.5, applied to L3 for an unbraced length equal to L1 or L2, as shown in Figure 4-2. If the structures are constructed using battered poles rather than parallel poles, the unbraced length for in-plane bending would be the greater of L1 or L2.

To determine the unbraced length factor for compression members of X-brace systems bending away from the plane of bracing, a different

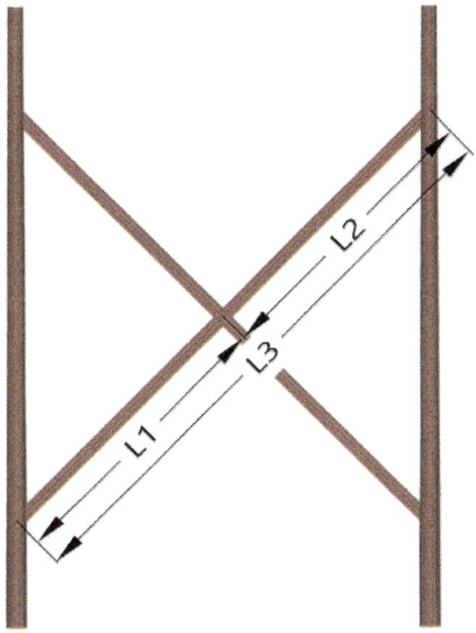

Figure 4-2. Tension/compression X-brace system.
Source: Otto Lynch, reprinted with permission.

approach must be followed. If the tension member of the X-brace has a force of not less than 20% of the force in the compression member, the crossover point connection provides some support that resists out-of-plane bending. ASCE 10 for lattice steel structures and ASCE 48-11 for tubular steel structures (ASCE/SEI 2012) both specify an unbraced length factor of K = 0.75, applied to L3 for out-of-plane bending when constructed with parallel poles. For structures with battered poles rather than parallel poles, the out-of-plane K factor is calculated as the greater of

$$K = (L1 + \tfrac{1}{2} L2) / L3 \tag{4-10}$$

or

$$K = (L2 + \tfrac{1}{2} L1) / L3 \tag{4-11}$$

This method of specifying the unbraced length factor for out-of-plane bending is consistent with the American Institute of Steel Construction (AISC) criteria in Specification Section C.

Other end-conditions and connections may warrant using a lower K factor. Engineering judgment combined with substantiated test data can be used to specify different K values than these recommendations. It should be noted that if the member in tension has a force of less than 20% of the force in the compression member, the crossover point may not provide adequate out-of-plane restraint and therefore the unbraced length of the member is the full length, thus $K = 1.0$ would be applied to L3.

4.3.3 Stress Equations

Round wood poles used for overhead line structures will be subject to the three primary stresses: bending, axial, and shear stresses. The basic formula for bending stress is

$$f_b = M_r / S \qquad (4\text{-}12)$$

where
f_b = Fiber stress in bending,
M_r = Sum of all applied primary and secondary bending moments acting on the cross-section being analyzed,
S = Section modulus of the cross-section being analyzed for round wood poles, which is calculated as

$$S = \pi D^3 / 32 \qquad (4\text{-}13)$$

where D is the pole diameter at the cross-section being analyzed.

The basic formula for axial stress is

$$f_a = P / A \qquad (4\text{-}14)$$

where
f_a = Stress on the section caused by axial load,
P = Sum of all axial loads acting on the cross-section being analyzed, and
A = Cross-sectional area at the location being analyzed for round wood poles, calculated as follows:

$$A = \pi D^2 / 4 \qquad (4\text{-}15)$$

where D is the pole diameter at the cross-section being analyzed.

The combined bending and axial stress equation is

$$f_b / F_b + f_a / F_a \leq 1.0 \qquad (4\text{-}16)$$

where
f_b = Fiber stress in bending,
F_b = Nominal strength in bending (MOR),
f_a = Stress on the section owing to axial load, and
F_a = Nominal strength in compression (generally MOR, unless reduced by buckling).

4.4 LOCALIZED BENDING CAPACITY REDUCTION BECAUSE OF BOLT HOLES OR OTHER DEFECTS

Drilled holes will reduce the strength of a wood pole by decreasing the area at a given cross-section, and thus the section modulus is decreased as well. A bolt hole oriented in the longitudinal direction of the spans and located in the neutral axis will cause very little reduction in the capacity of the pole to support the wind loads on the wires at that cross-section. If the hole is oriented in the transverse direction, perpendicular to the spans, there will be a reduction in the section modulus and pole capacity at that cross-section for supporting the wind loads. To calculate the reduction in section modulus for bolt holes in the most common directions, shown in Figure 4-3, the following formulas can be used.

Reduction in cross-section area is

$$D \times b \tag{4-17}$$

where D is the diameter of the wood pole at the cross-section with the bolt hole, and b is the bolt hole diameter (in.).

Reduction in transverse section modulus for a bolt hole in the transverse direction is calculated as

$$S_t = (\pi D^3/32) - (b \times D^2/6) \tag{4-18}$$

Reduction in transverse section modulus for a bolt hole in the longitudinal direction is calculated as

$$S_l = (\pi D^3/32) - (D \times b^2/6) \tag{4-19}$$

For bolt holes in oblique directions, a more rigorous method of calculating the section modulus can be used.

4.5 RESILIENCY

Resiliency describes the overall robustness and recovery characteristics of overhead line infrastructure. Greater resiliency helps to avoid or minimize interruptions of service and restoration time during and following a major storm. Regarding structural design and analysis, it has been common practice to design structures individually rather than consider the interaction of structures within the line system. However, software tools are rapidly advancing to be able to effectively and efficiently perform an analysis from "point A to point B" of a line. Engineering judgment should determine how to use such software tools.

Effective maintenance programs help preserve the strength and resiliency of wood poles. Fewer weakened poles means fewer pole failures and outages, shorter restoration times, and lower restoration costs. The concept

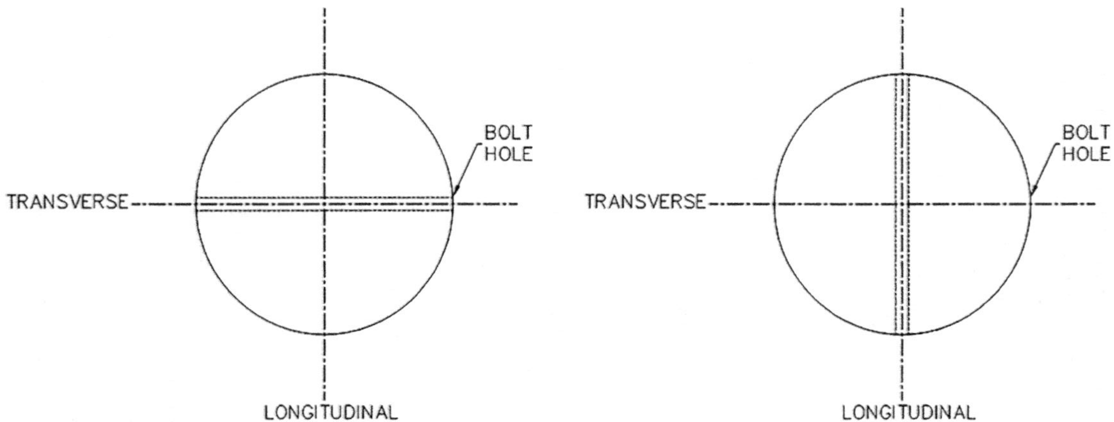

Figure 4-3. Orientation of bolt holes.
Source: Vicki Schneider, reprinted with permission.

of resilience is discussed in greater detail in Chapter 9, "Inspection, Assessment, Maintenance, and Repair," and in Appendix A.

REFERENCES

ANSI/AWPA (American Wood Protection Association). 2001. "Evaluation of fiber stress values for wood utility poles." In ANSI-O5.1, *Wood poles—specifications and dimensions*. Birmingham, AL: AWPA.

ANSI/AWPA. 2017. ANSI-O5.1. *Wood poles—specifications and dimensions*. Birmingham, AL: AWPA.

ASCE. 2010. *Guidelines for electrical transmission line structural loading*, MOP 74. Reston, VA: ASCE.

ASCE. 2012. *Prestressed concrete transmission pole structures recommended practice for design and installation*, MOP 123. Reston, VA: ASCE.

ASCE/SEI. 2006. *Reliability-based design of utility pole structures*, MOP 111. Reston, VA: ASCE.

ASCE/SEI. 2012. Standard 48-11, *Design of steel transmission pole structures*. Reston, VA: ASCE.

ASCE/SEI. 2015. Standard 10-15, *Design of latticed steel transmission structures*. Reston, VA: ASCE.

Bohannan, B., H. Habermann, and J. E. Lengel. 1974. "Taper of wood poles." *Gen. Tech. Rep. FPL* 2. Madison, WI: USDA Forest Service, Forest Products Laboratory.

California PUC (Public Utilities Commission). 2018. *Rules for overhead line construction*, General Order No. 95. Accessed November 18, 2018. http://www.cpuc.ca.gov/gos/GO95/go_95_startup_page.html.

IEEE (Institute of Electrical and Electronic Engineers). 2017. ANSI Standard C2, *National electrical safety code*. New York: IEEE.

CHAPTER 5
CONNECTIONS

5.1 GENERAL

The physical attachment of arms, brackets, insulators, and climbing provisions to wood poles is a very important consideration to the overall structure design. These connections serve the explicit purpose of transferring load to the supporting pole or frame structure. Pole line hardware is attached to the pole using bolts placed in through holes drilled in the pole cross-section. Pole bands wrapping the pole have also been used depending on the type and magnitude of the load. Factors to be considered in connection design include the load transfer mechanism, ductility, durability, required pole and bracket fabrication tolerances, aesthetics, constructability, and general economics.

Attachments need to be properly designed to ensure proper load transfer, and the connections should be developed by an engineer experienced in wood pole structure design. In some instances, the pole line hardware or framing supplier may also be able to provide valuable assistance with typical connection design details. The structural adequacy of connections can be determined empirically through full-scale component testing or by using accepted engineering practices. The connections should be designed such that the design strength in both the connecting part and the wood pole is not exceeded and excessive deformation or connection rotation does not occur.

Typical framing materials and attachments to wood pole structures are presented in Figures 5-1 through 5-8.

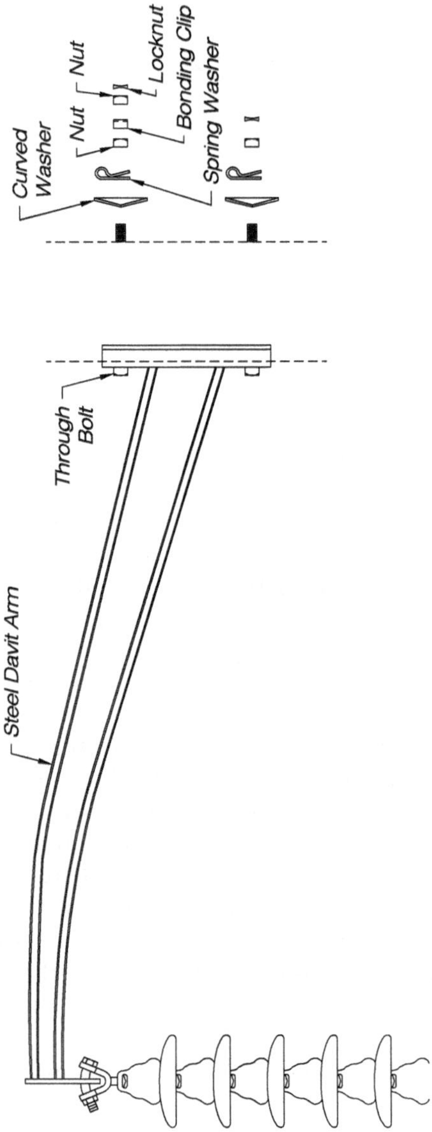

Figure 5-1. Davit arm connection.
Source: Hughes Brothers, reprinted with permission.

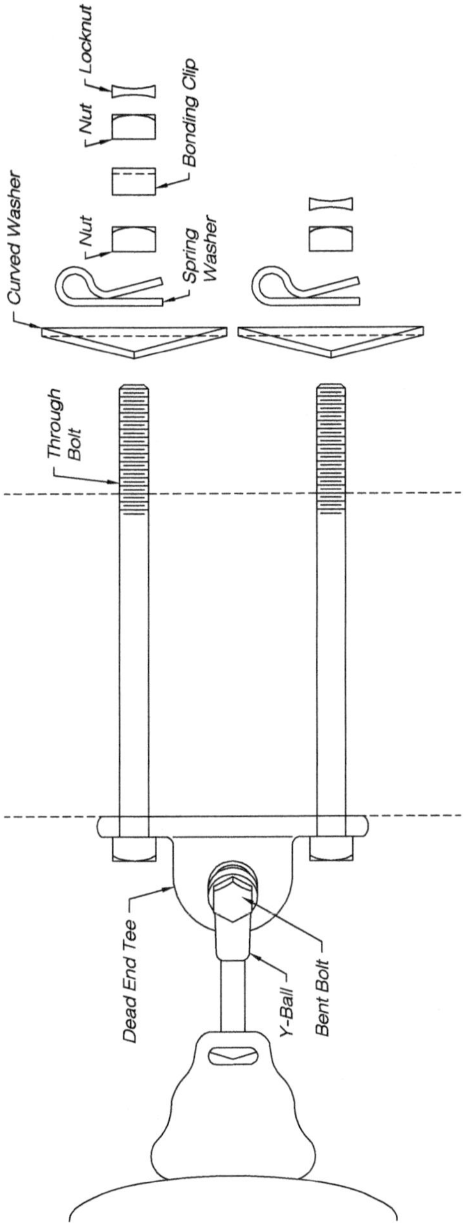

Figure 5-2. Dead-end insulator connection.
Source: Hughes Brothers, reprinted with permission.

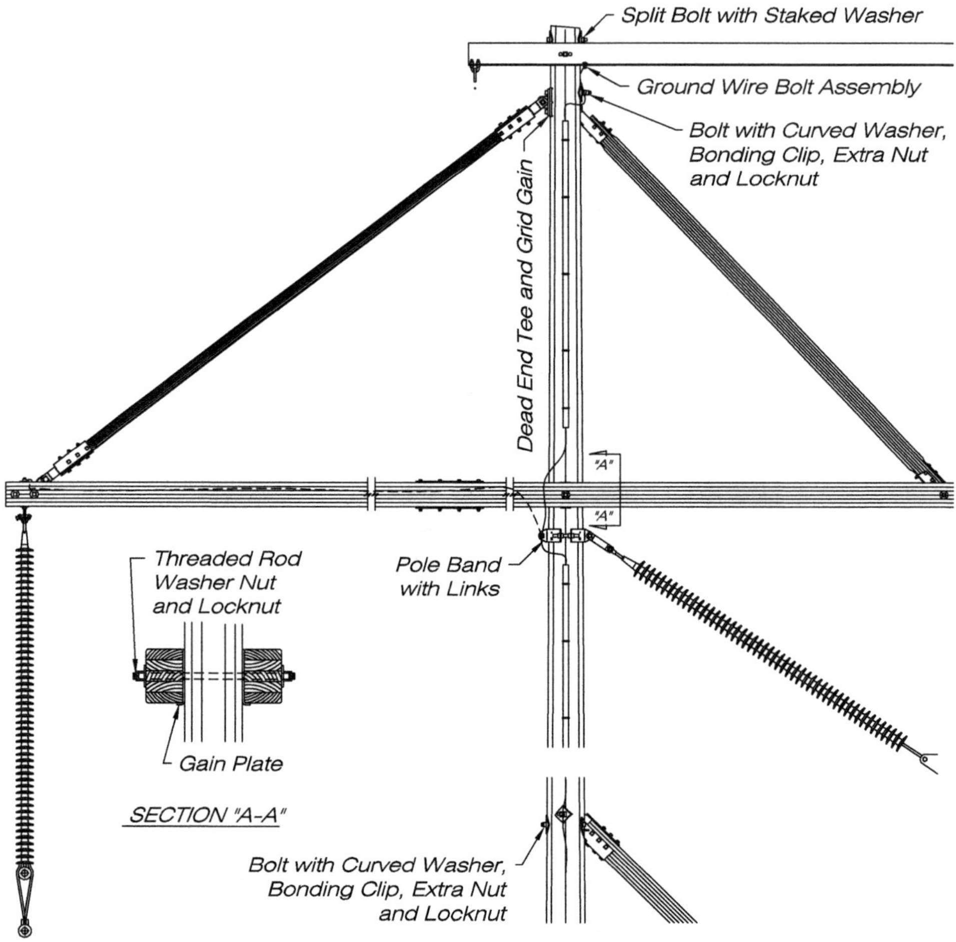

Figure 5-3. H-frame crossarm, vee brace, and knee brace.
Source: Hughes Brothers, reprinted with permission.

5.2 THROUGH BOLTS

Most pole line hardware is connected to wood poles using galvanized through bolts. Commonly used fastener standards include ANSI C135.1 (IEEE 1999) and C135.80 (IEEE 2012), ASTM A307 (ASTM 2014a), ASTM A193B7 (ASTM 2017), and ASTM A449 (ASTM 2014c). A common misconception is that A325 bolts (ASTM 2014b) are used, but A325 bolts have a strict thread length per ASTM and are nearly always too short for wood pole applications; therefore, A449, which is identical in all other ways, should be specified.

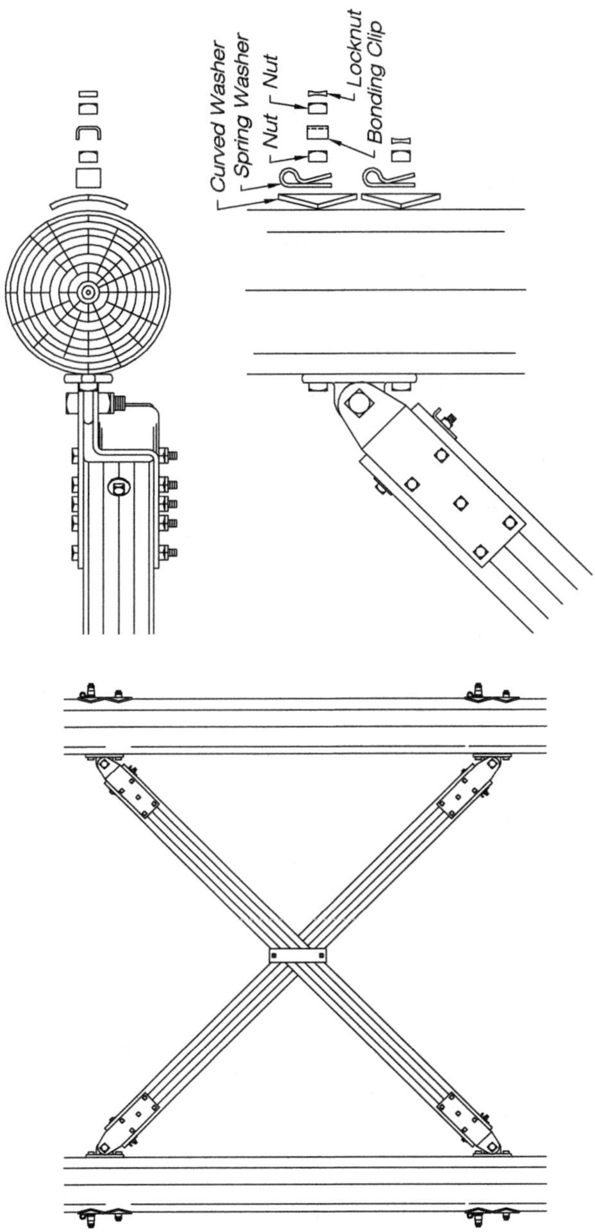

Figure 5-4. X-bracing assembly connection.
Source: Hughes Brothers, reprinted with permission.

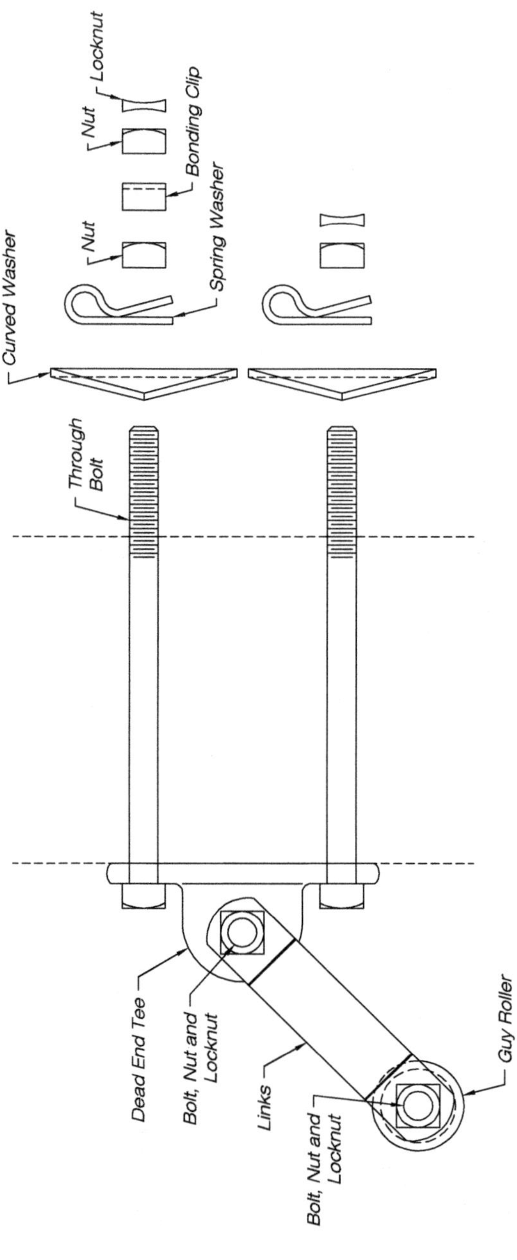

Figure 5-5. Guying connection.
Source: Hughes Brothers, reprinted with permission.

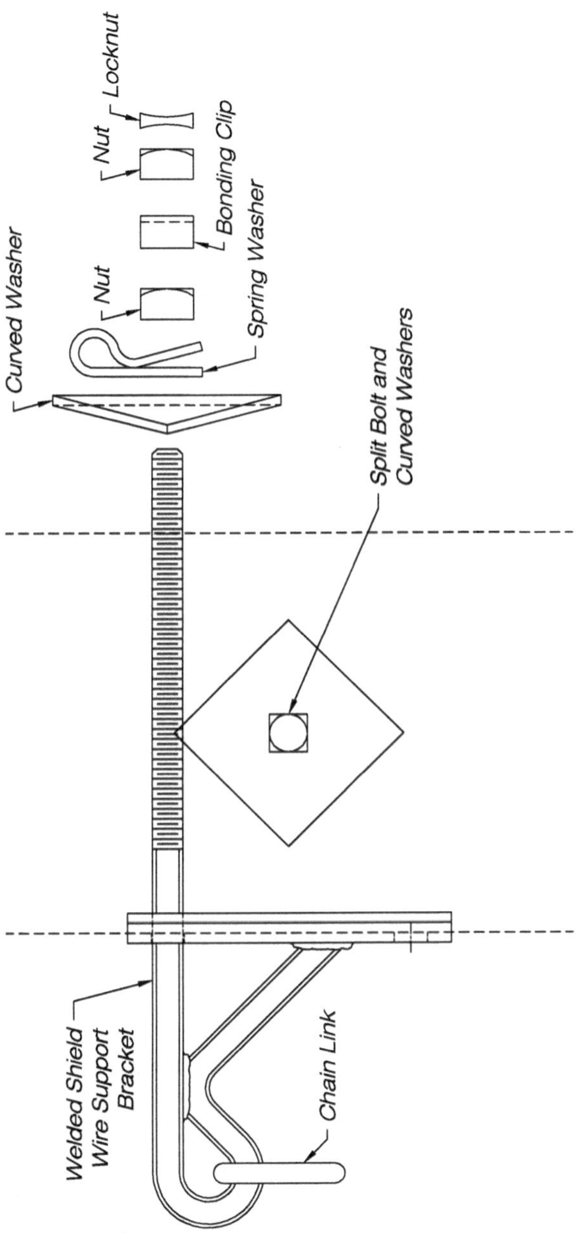

Figure 5-6. Shield wire attachment assembly connection.
Source: Hughes Brothers, reprinted with permission.

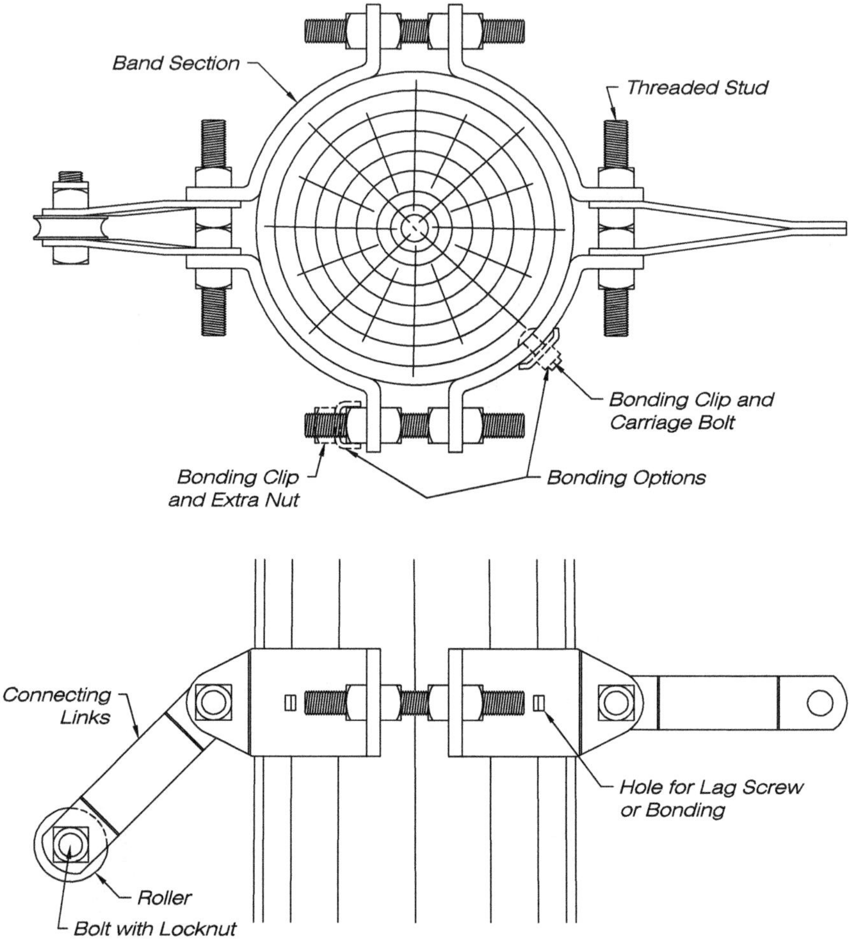

Figure 5-7. Formed banding connection.
Source: Hughes Brothers, reprinted with permission.

The engineer responsible for the connection design should determine the fastener standard, the type selected, and the bolt-tightening procedure used. There are several references available regarding proper torque values for connections involving strictly steel-steel components, but these values do not apply to joints of soft materials such as wood poles. Typically, the force used to tighten a nut with a 12 in. wrench until complete resistance is met is considered sufficient.

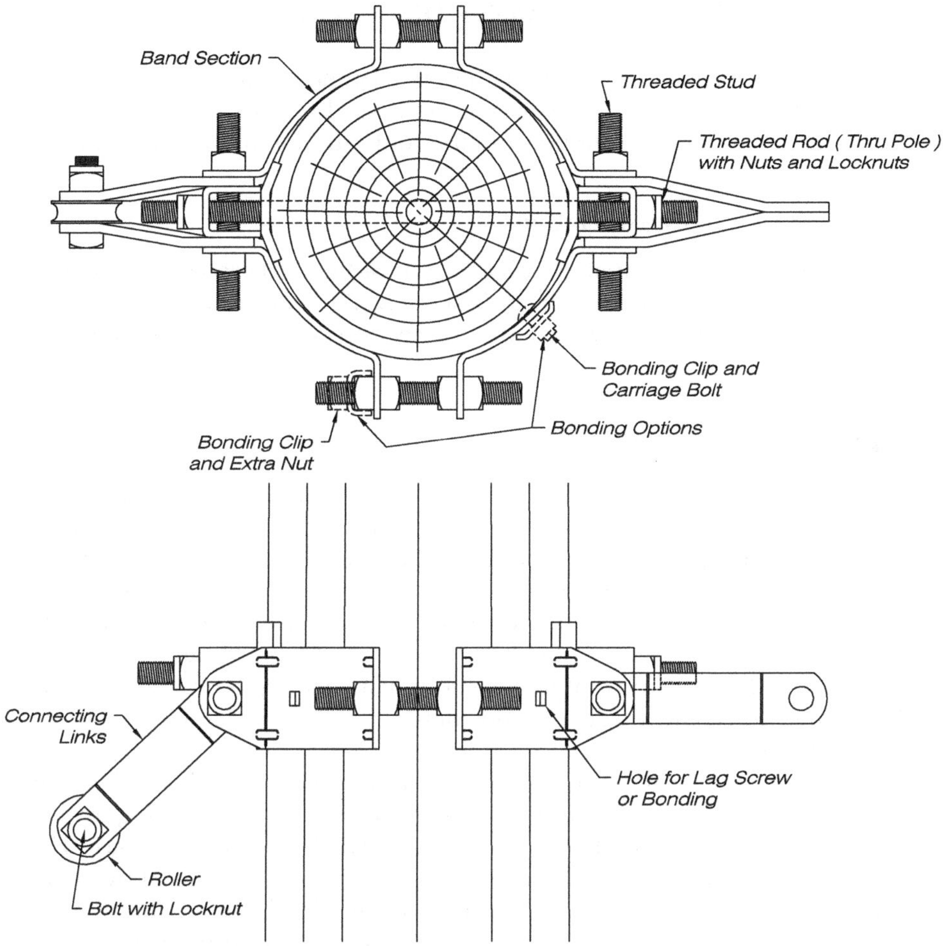

Figure 5-8. Formed banding with through-bolt connection. Source: Hughes Brothers, reprinted with permission.

5.3 BANDED CONNECTIONS

Round wood poles are well suited for attaching hardware by banding using formed steel bands, steel strapping, or cast steel chain links. Banding systems must be designed for the loads they will support. All banding alternatives are adjustable for the various ranges of pole diameters. Some use a bolt through the pole to support higher loads.

5.4 CONNECTION DESIGN CONSIDERATIONS

During the design of connections, the engineer needs to consider eccentric loads and torsional moments, combined stresses, localized stress concentrations, and other loads resulting from the configuration of the connection.

5.4.1 Congestion Considerations

Careful attention must be given to avoiding excessive attachment details in a localized area of the pole section. This congestion reduces the cross-section and may decrease the connection capacity and weaken the pole.

5.4.2 Bearing Failure Considerations

Careful attention should be given to pole surface-bearing stress. Surface-bearing stress refers to the load distribution at the pole surface-attachment interface. Examples include the concentrated load underneath the washer of a through bolt or the contact surface on the compression end of a pole eye plate. A curved washer should be placed between the pole and the head and/or nut of the bolt to adequately distribute the concentrated load. The thickness of the washer must be adequately sized for the load and the bearing area being considered. The use of cast washers is not recommended.

The addition of grid gains (also known as pole gaining), shear plates, or gain plates can lessen pole surface-bearing stress. Grid gains (see Figure 5-9) are spiked connectors that do not require the creation of a flat surface. The spikes or teeth penetrate the wood surface of the pole and resist movement with and against the grain. The gains distribute loads over a larger area, strengthening the joint and reducing fiber crushing and hole slotting as seen in Figure 5-10. Gains are especially useful when the mounting bolt is subjected to large shear loading (note the elongation of the hole and wood fiber crushing in Figure 5-11). Gains have the additional advantage of creating a steel-wood connection that separates wood surfaces, a possible source of premature decay.

5.4.3 Banding Systems

Banding systems are designed based on the magnitude of the load and the direction from which it is applied. The engineer should be knowledgeable in selecting the appropriate banding system to withstand the load and should verify with the supplier that the strength of the systems is adequate for the design load it will support.

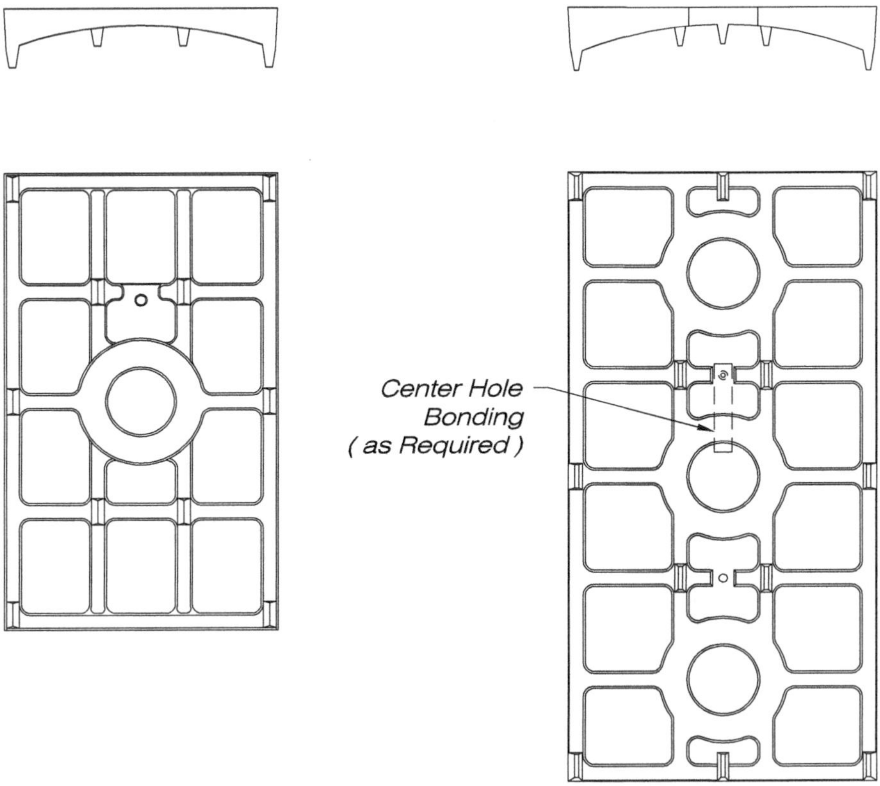

Figure 5-9. Toothed grid gains.
Source: Hughes Brothers, reprinted with permission.

Pole bands with through-bolt support provide more vertical load strength (see Figure 5-12 and Figure 5-13) and limit downward movement in guying applications. Figure 5-14 shows that the load on the guy-side of the band controls the banding system design and can be estimated using simple geometry:

$$T = \frac{P}{sin\phi} \tag{5-1}$$

where
T = Tension in guy wire, in pounds (lb),
P = Horizontal load applied to the band, in lb, and
Ø = Angle between guy wire and pole.

Figure 5-10. Pole after loading with grid gains.
Source: Hughes Brothers, reprinted with permission.

Figure 5-11. Pole after loading without grid gains.
Source: Hughes Brothers, reprinted with permission.

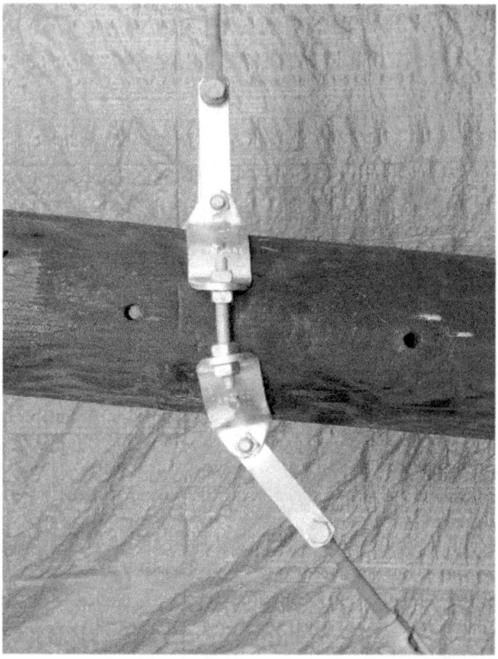

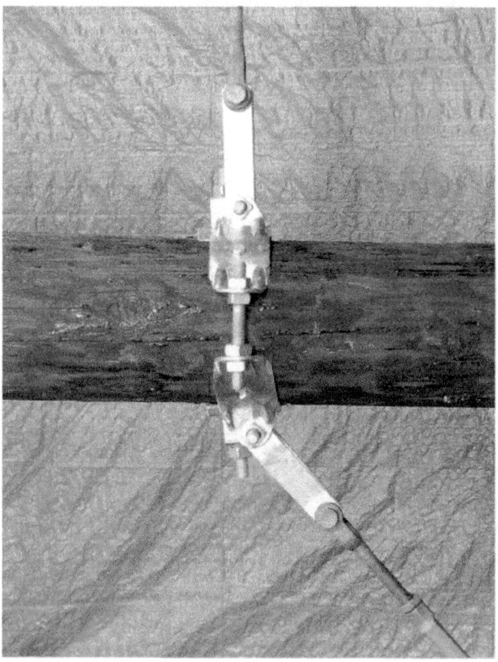

Figure 5-12 (left). Bands with through bolts.

Figure 5-13 (right). Bands without through bolts.
Source: Hughes Brothers, reprinted with permission.

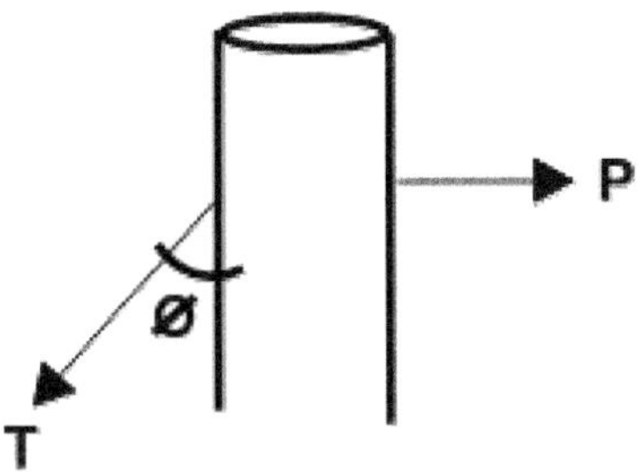

Figure 5-14. Geometry of a guyed structure.
Source: Vicki Schneider, reprinted with permission.

5.5 CONNECTION FAILURE MODES

Failure of connections may occur in the wood fibers or the metal components. This section explains several of the particular failure modes. Refer to National Design Specification (NDS) Supplement, "Design Values for Wood Construction," (NDS 2018) for values necessary for calculations related to connection failures.

5.5.1 Fiber Crushing Perpendicular to Grain

Crushing of pole fibers perpendicular to the grain can occur when the surface-bearing area of the connection or washer is not sufficient to distribute the load or when the connecting part is not of sufficient thickness to prevent deformation of the part (see Figures 5-15 and 5-16). The resulting fiber crushing, while not necessarily the sign of eminent failure, should be avoided.

5.5.2 Fiber Crushing Parallel to Grain

Fiber-bearing failure parallel to grain occurs when the bearing capacity of the wood fibers is exceeded, which results in crushing of the pole. The resulting failure can cause hardware to loosen or deform; in extreme cases it can cause the pole to split.

The maximum bolt-bearing load can be estimated by multiplying the diameter of the bolt times the diameter of the pole times the fiber compressive strength parallel to grain. An L/d factor is then applied since a longer bolt will tend to bend, crushing outer fibers first (see Figure 5-17). The calculation is as follows:

$$P = fd(L)(\emptyset) \tag{5-2}$$

where
P = Maximum load, in pounds;

Figures 5-15 (left). 1/4 in. washer.

Figure 5-16 (right). 3/8 in. washer under same load.
Source: Hughes Brothers, reprinted with permission.

f = Fiber compression strength parallel to grain;
d = Bolt diameter, in inches;
L = Length of embedded bolt, in inches (equal to pole diameter); and
Ø = 0.35 (L/d adjustment factor for wood poles).

5.5.3 Bolt Shear or Tension Failure

Bolt shear or tension failure occurs when the allowable shear or tensile capacity of the bolt is exceeded. Bolts should be selected based on the design shear or tension tabulated in the appropriate standards such as ASTM A307 (2014a) or IEEE C135.1 (IEEE 1999).

The designer shall take into consideration that shear strengths through the threads are different from through the shaft.

For bolts that do not have a specified shear or tensile strength, the following equation shall be used:

$$F_v = 0.62F_u \tag{5-3}$$

where F_v is the design shear strength on the effective area, and F_u is the specified minimum tensile strength of the bolt material.

The effective area is the gross cross-sectional area of the bolt if the threads are excluded from the shear plane or the root area if the threads are in the shear plane. For bolts having a specified proof-load tensile stress, the tensile capacity is equal to the ASTM proof-load stress obtained by the length

Figure 5-17. Bolt and hardware bearing on pole with excessive fiber crushing evident.
Source: Hughes Brothers, reprinted with permission.

measurement method. For bolts with no specified proof-load stress, the tensile capacity is equal to

$$F_t = 0.6F_u \quad (5\text{-}4)$$

where F_t is the design tensile strength of the effective area, and F_u is the specified minimum tensile strength of the bolt material. Figures 5-18 and 5-19 show two bolt failures.

5.5.4 Bolts Subject to Combined Shear and Tension

For bolts subject to combined shear and tension, refer to the appropriate sections of the AISC Steel Construction Manual (AISC 2017) as well as USDA Technical Bulletin 332 (USDA 1932). The interaction equation in these documents is a good method to estimate bolt stresses in combined loading conditions.

Figures 5-18 (left). Failure of bolt in threads.

Figure 5-19 (right). Failure of bolt in shaft.
Source: Hughes Brothers, reprinted with permission.

5.5.5 Bolt Bending or Prying Action

When designing a connection, special consideration should be given to the possibility that bending of the bolt and/or a prying action can result in bolt failure.

5.6 INSTALLATION CONSIDERATIONS

The manufacturer's instructions regarding bolt tightening should be followed for all attachment connections. The engineer should be aware that overtightening of the bolts may lead to fiber crushing or hardware deformation.

Bolts and hardware on structures should be grounded to eliminate the potential for crossarm or pole fires because of leakage current. Common ground wire bonding arrangements to steel and to wood are shown in Figure 5-20 and Figure 5-21, respectively.

Wood framing products and poles tend to shrink over time as they lose moisture content and as the wood weathers under wind and UV exposure. Connection hardware should be kept static-free and tight by using tapped hardware fittings and spring washers. Examples of spring washers are shown in Figures 5-22 and 5-23.

Figure 5-20. Ground wire bonding to steel.
Source: Hughes Brothers, reprinted with permission.

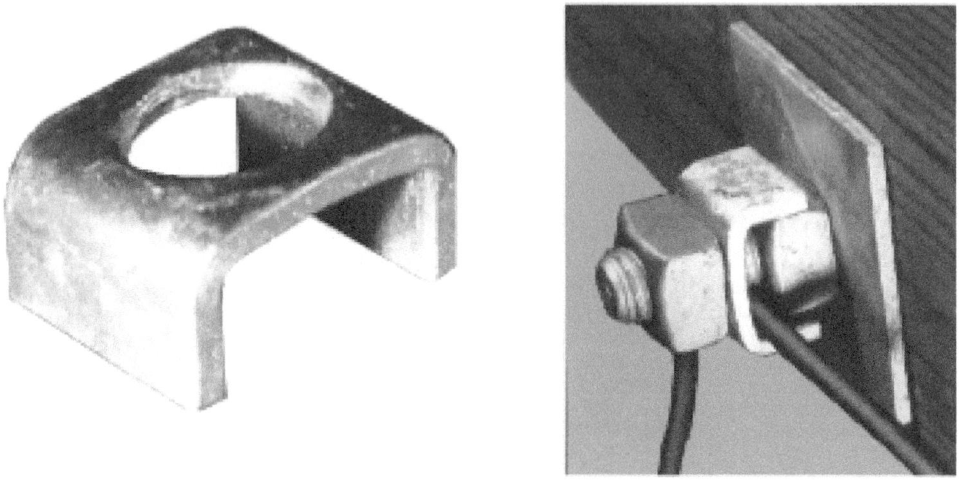

Figure 5-21. Ground wire bonding to wood.
Source: Hughes Brothers, reprinted with permission.

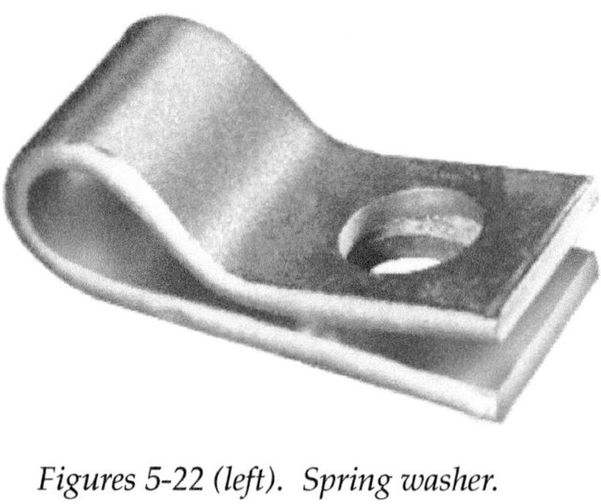

Figures 5-22 (left). Spring washer.

Figure 5-23 (right). Double coil lock washer.
Source: Hughes Brothers, reprinted with permission.

REFERENCES

AISC (American Institute of Steel Construction). 2017. Steel Construction Manual, Fifteenth Edition. United States of America.

ASTM International. 2014a. Standard A307, *Standard specification for carbon steel bolts, studs, and threaded rod 60 000 psi tensile strength*. West Conshohocken, PA: ASTM.

ASTM. 2014b. Standard A325, *Standard specification for structural bolts, steel, heat treated, 120/105 ksi minimum tensile strength*. West Conshohocken, PA: ASTM.

ASTM. 2014c. Standard A449, *Standard specification for hex cap screws, bolts and studs, steel, heat treated, 120/105/90 ksi minimum tensile strength, general use*. West Conshohocken, PA: ASTM.

ASTM. 2017. Standard A193, *Standard specification for alloy-steel and stainless steel bolting for high temperature or high pressure service and other special purpose applications*. West Conshohocken, PA: ASTM.

IEEE (Institute of Electrical and Electronic Engineers). 1999. Standard C135.1, *Standard for zinc-coated steel bolts and nuts for overhead line construction*. New York: IEEE.

IEEE. 2012. Standard C135.80, *Standard for fasteners for overhead line construction*. New York IEEE.

NDS (National Design Specification) Supplement. 2018. *Design Values for Wood Construction*. Leesburg, VA: American Wood Council.

USDA (US Department of Agriculture). 1932. "The bearing strength of wood under bolts." *Tech. Bull. 332*. Washington, DC: Government Printing Office.

CHAPTER 6

WOOD POLE FOUNDATIONS AND ANCHORING

6.1 GENERAL

A wide variety of wood pole structures can support overhead lines of varying voltages. Wood poles are generally direct embedded, so the foundation is part of the pole, yielding a cost-effective means of supporting an overhead line. As with other pole types, it is essential to properly design and install foundations for the long-term stability of wood pole structures.

Direct-embedded wood pole structures (see Figure 6-1) must resist vertical, transverse, and longitudinal loads in a way that efficiently distributes the loadings to be resisted by the earth. The embedded portion of a pole must be designed such that serviceability requirements of the line are maintained under the required code and design loading conditions. For a single self-supported pole, these forces could include overturning moments, torsion, and horizontal and vertical forces. In addition, dead-end and angle structures use guy wires to reduce differential longitudinal or transverse loadings on the structure, resulting in lower overturning moments at the base of the structure.

The engineer should take into account soil types and soil characteristics during the design of embedment lengths. The engineer should also design to limit displacements and/or settlements to prevent any adverse effect to the reliability and service of the transmission line.

6.1.1 Performance and Reliability

Foundation design typically considers the performance and reliability requirements for the transmission structure, which are defined by the allowed rotation and deflection at the groundline of the structure.

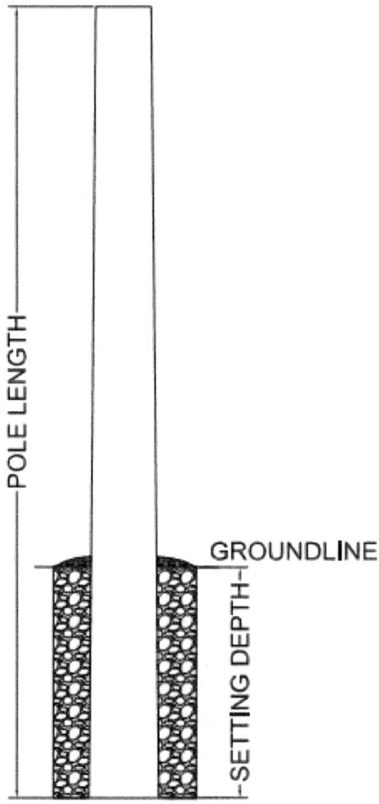

Figure 6-1. Direct-embedded pole.
Source: Vicki Schneider, reprinted with permission.

The deflection limit is typically described by recoverable deflection and nonrecoverable deflection. It is the displacement measured at the groundline. Recoverable deflection is the distance that the foundation will displace and return to a nondisplaced position after the load has been removed. Nonrecoverable deflection is the displacement that is permanent after the load is removed.

Rotation of the foundation is measured at the point of fixity typically either at the base of the pole or at two-thirds of the embedment depth. The amount of rotation on the foundation affects the pole-top displacement and can lead to larger-than-expected deflections, reduced horizontal clearances, and increased secondary moments.

Deflection and rotation limits should be selected based on the type of structure, the allowable deflection/displacement of attachments on the structure, pole-top deflection, aesthetics, and pole strength limits. The engineer may want to consider the nonlinear effects of the allowed deflection and rotation of the foundation on the structural capacity of the pole.

Deflection and rotation limits are typically a better means of controlling foundation performance compared to using arbitrary soil safety factors.

6.1.2 Historical Perspective on "Rule of Thumb"

In the past, many utilities adopted a "rule of thumb" standard bury depth of 10% of the pole length plus 2 ft. This bury depth was based on the approximate groundline locations provided in Tables 3 through 10 of ANSI-O5.1 (ANSI/AWPA 2017) covering dimensions of poles. These approximate groundline locations generally, but not always, follow the 10% of pole length plus 2 ft rule-of-thumb standard, except for short poles and poles longer than 90 ft. However, those locations were never intended to define embedment depths representing adequate foundation capacity for design/construction purposes.

The approximate groundline locations had two purposes: first, to serve as a reference location for defects that may or may not be allowed in the groundline zone; and second, to establish a setting depth where the designated fiber strength in ANSI-O5.1 Table 1 and the minimum pole circumference values at groundline combine to support the horizontal tip loads in ANSI-O5.1 Annex B applied 2 ft from the tip. The required circumference at the approximate groundline is established for each species to account for differences in fiber strength. Thereby, poles of the same length and class of different species have similar load capacity.

Although many have used the ANSI-O5.1 approximate groundline locations or the rule-of-thumb embedment of 10% of the pole length plus 2 ft, foundation failures have occurred, especially with larger class poles, as shown in Figure 6-2. Others use various rules of thumb to determine embedment depth, but these approaches usually do not adequately address the actual soil conditions. These basic rule-of-thumb methods also do not address bearing capacity for vertical forces caused by frame or guying action. A growing trend in the industry is for transmission companies to abandon the rule of thumb and determine required embedment depths considering actual loading and soil conditions at specific structure locations.

Several studies evaluating the rule-of-thumb method have shown that in many cases, this method does not provide adequate embedment depth for heavily loaded poles or certain soil conditions. (For more information on wood pole bury depths in various soils, see Lutein 2009.)

Embedment depths should be calculated based on project- or location-specific soil conditions, performance and reliability requirements, and expected loading conditions as discussed in Section 6.2.

Figure 6-2. Foundation failure.
Source: Jaron Reay, reprinted with permission.

6.2 DESIGN CONSIDERATIONS

During the foundation design process, the engineer should identify the maximum pole groundline forces and apply them to the soil strength properties. A proper foundation design is achieved when the performance requirements for strength and serviceability are met. Foundation designs should provide adequate support for all dead and live loads, horizontal shear, bending moment, vertical uplift, and vertical compression loads. The foundation engineer must be certain that the loads from the structure in each design load condition and the allowable loads for the soils are expressed in either working loads or ultimate loads. When using working loads, a factor of safety is specified to ensure an adequate level of reliability is achieved. When using ultimate loads, a load factor is applied to meet the design requirements.

6.2.1 Soil Exploration and Design Parameters

Soil testing and geotechnical analysis may be required to determine the conditions and characteristics of the soil and the type and size of foundations that will be required. Augered borings or driven casings are most common in soil types other than bedrock. Core samples are taken in bedrock areas. The report developed from the soil tests should include the bearing capacity, lateral soil strength, the cohesion values, and the depth of the water table. The engineer shall specify the locations along the transmission line where soil testing shall be performed, depending on the local geology, structure loading conditions, and overall reliability desired for the foundation designs. The greater the distance between the soil sample locations, the more conservatively the foundations should be designed.

The engineer could elect not to perform soil testing if geological information is available or if the soil conditions are considered consistent. If the engineer elects not to take borings, a field survey could be performed to obtain general soil classification information. A field survey could include the use of a hand auger, penetrometer, or torque probe to determine general soil conditions. However, if the poles are heavily loaded angle or dead-end structures or the soil conditions are unknown, performing soil testing is recommended.

Design Manual for High-Voltage Transmission Lines (USDA 2015), provides a table to assess soil classification. The two main soil descriptions used are sand (noncohesive) and clay (cohesive). Other common materials encountered are rock and organic material such as peat and coal. The engineer should seek specific geotechnical advice and recommendations regarding the foundation designs when working in areas with rock and organic material.

In general, a soft soil will provide less capacity than hard or stiff soil, but clay and sand will affect bearing, moment, uplift, and shear capacities differently, so each should be evaluated independently. Constructability should also be considered because loose sand and soft clays may need to be cased when augered to prevent the excavation from collapsing during construction.

Side slopes should also be noted and considered in the foundation design if the slope is such that additional embedment is required to develop the full strength of the soil on the downhill side.

6.2.2 Foundation Reactions

Calculation of foundation reactions should follow a basic theory of statics. The engineer should establish a coordinate system in relation to the transmission line's alignment and all transverse, longitudinal, and vertical forces; and overturning moments shall be calculated at the groundline per the established coordinate system.

Guidelines for Electrical Transmission Line Structural Loading (ASCE 2010) provides loading methodologies based on Load and Resistance Factor Design (LRFD). The general format of the LRFD equation is that the strength ($\emptyset R$) is equal to or greater than the accumulated factored loads ($\Sigma\gamma Q$).

$$\emptyset R \geq \Sigma\gamma Q \tag{6-1}$$

where

$\emptyset$ = Strength factor (phi),

R = Measure of resistance (strength),

γ = Load factor, and

Q = Calculated load.

The foundation reactions are calculated using the applicable required loads per *Guidelines for Electrical Transmission Line Structural Loading* (ASCE 2010) loading methodology.

6.2.3 Framed Structure Embedment Considerations

A cross-braced framed structure will transfer much of the transverse loads to vertical uplift and compressive loads owing to the rigidity of the structure. Therefore, the engineer must consider bearing and uplift in the design of the structure foundations. Ground slope must also be considered as the minimum required embedment depth must be achieved on all poles of the framed structure.

6.2.4 Foundation Design and Analysis Software

A wide variety of software programs are used to analyze and design utility-type foundations. These programs use a finite element analysis and matrix methods to calculate deflections, settlements, bearing capacity, lateral earth pressures, stresses, shear, and bending moments. If engineers elect to use a software program to design foundations, they should have a basic understanding of the design theory or method on which the software is based, and understand its capabilities and/or limitations. A combination of hand calculations, sound engineering judgment, and a history of wood pole embedment should be used to validate the engineering software designs.

Each design method requires a different set of input parameters, but the minimum items necessary are depth of each soil layer, density of soil, shear strength for cohesive soil, friction angle for noncohesive soil, and the depth of the water table. The geotechnical parameters required for the particular method or the software used in design should be examined before a geotechnical investigation is conducted.

6.3 FOUNDATION TYPES

Foundations for wood pole structures can typically be classified in one of the following ways: as direct embedments with either native or engineered backfill, or special foundations that have been engineered for a specific site.

6.3.1 Direct Embedment and Backfills

Direct embedment of wood pole structures is typically the foundation of choice, based on its low cost and ease of installation in moderate to good soil conditions.

Direct embedment as shown in Figure 6-3 consists of augering a hole, inserting the wood pole, and backfilling (see Figure 6-4). The engineer is responsible for specifying the required type of backfill, which can consist of compacted native soil or engineered backfill including compacted sand, gravel, crushed rock, or expanding foam. Special consideration is needed when using native soils during cold weather construction. Depending on the soil type, the hole may require the use of a casing to prevent cave-ins during construction.

Direct embedment can also consist of jetting the wood pole into the soil with hydrostatic pressure. As the native soil is hydraulically displaced, the self-weight of the wood pole structure forces the embedment of the pole. Jetting is useful in granular soils with a high water table in which an

Figure 6-3. Direct-embedded pole.
Source: Justin Johnson, reprinted with permission.

Figure 6-4. Compaction of backfill.
Source: Justin Johnson, reprinted with permission.

augered hole tends to cave-in. Jetting, however, can loosen the native soils, which may temporarily compromise in situ soil properties. Consequently, jetting should only be used in granular soils, preferably uniformly graded sands without organic material.

6.3.2 Special Foundations

In areas with poor soil conditions that make direct embedment impractical or inadequate, other foundation options should be considered. Poor soil conditions include organics, soft clay, and loose sand. Several alternative foundation design options are listed in the following sections. If any of the options are deemed practical for the structure in question, then the engineer should coordinate with the contractor or owner's construction coordinator to determine which option is preferred from a constructability standpoint.

6.3.2.1 Side Guy Foundation System. The side guy foundation system requires the installation of two or four anchor guys to aid the direct-embedded structure foundation in resisting the overturning moment. Refer to Figure 6-5. Side guys are installed primarily to reduce the overturning moment of the structure. The side guy method discussed here is not installed because of an insufficiency in the structural capacity of the wood pole or frame structure.

The minimum embedment depth (E) of these structures is the depth of the poor soil (L1) plus a calculated depth of acceptable soil (L2). The depth of L2 is to prevent the butt of the pole from kicking out or walking. This minimum depth should be calculated based on loads and soil strength properties. This method requires a review of the easements to determine if guying is allowed. For single wood pole structures as shown in Figure 6-6, side guys are in the transverse direction perpendicular to the wires.

6.3.2.2 Deep Set in Culvert. The deep set culvert foundation type consists of installing a culvert through the poor soil and into the acceptable soil to a depth specified by the engineer, as shown in Figure 6-7. If soil conditions allow, spiral culverts are preferred so that they can be advanced into the soil ahead of the excavation to reduce soil disturbances. This may require telescoping culvert sections.

The minimum embedment depth of these structures is the depth of the poor soil plus the calculated setting depth for the pole height that is being installed. The culvert is then backfilled with crushed rock or other backfill material specified by the engineer. If the bottom of the excavation is disturbed, a bearing pad of calculated thickness and material based on bearing capacity should be placed in the hole. Culverts should be sized so that they provide enough room for the selected backfill and compaction methods being used. In agricultural areas, the top of the culvert should be installed at least 2 ft below existing grade to reduce the probability of damage from farm equipment. In areas affected by standing water, the top of the culvert should be specified as 2 ft above the expected high water level, depending on the owner's safety requirements. The culvert can be set flush with existing grade for most other situations.

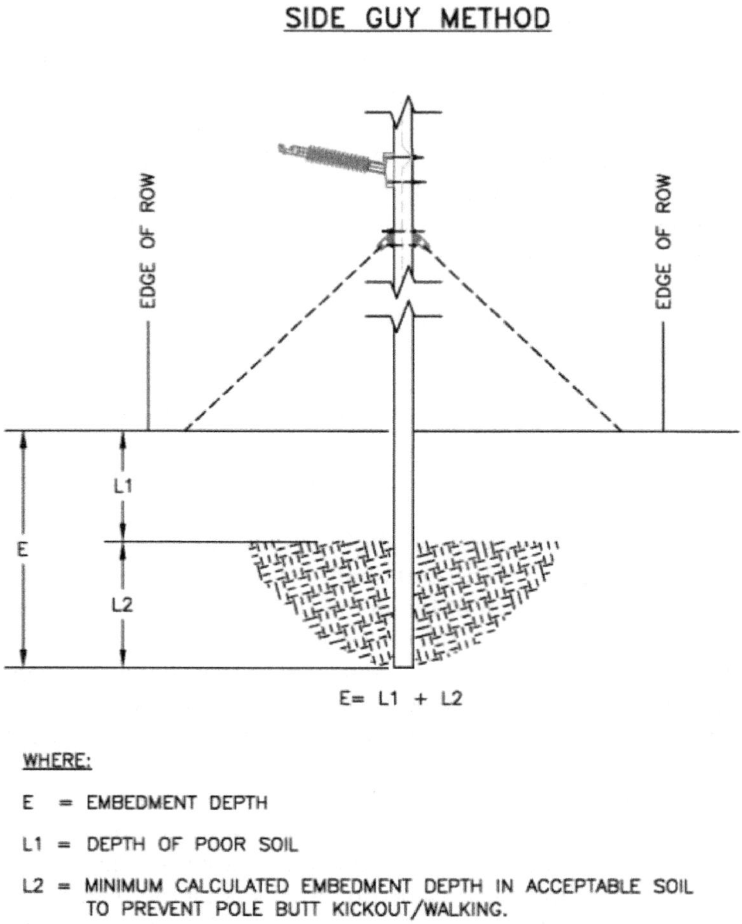

Figure 6-5. Side guy foundation system design.
Source: Great River Energy, reprinted with permission.

6.3.2.3 Pole Keys. The pole key is installed on the side of the embedded portion of the pole (see Figure 6-8) to increase the bearing area of the foundation against the soil and increase the overall capacity to resist lateral loads and overturning moments. Pole keys can be used for lightly loaded small poles in soft soil conditions. They should be installed below the frost depth to prevent damage to the pole key and maintain its effectiveness.

6.3.2.4 Screw Piles. Screw piles are used in deeper soft soils, organic soil, muskeg, and waterlogged areas where typical direct embedment of wood poles is not feasible. Screw piles are installed to pass through the soft

Figure 6-6. Typical side guy foundation system installation.
Source: Great River Energy, reprinted with permission.

soils until they reach the stronger soils below, as determined by the amount of torque required to advance the pile.

Screw piles are made with a steel tube with a diameter large enough to insert the wood pole. Below the steel tube is a second tube with multiple helices. The diameter of the second tube could be the same as or smaller than the pole tube. The length of the lower tube and number of helices depends on the depth the screw pile must be advanced to contact the strong soil layer. A detailed geotechnical exploration is required at each specific structure location to properly design the screw piles.

There are two screw pile types. Type-1 has a bottom tube with smaller tube diameter. Shown in Figure 6-9, this type of screw pile is used when good soils are available at a relatively shallow depth. Type-2 is a screw pile with uniform tubular section diameters. This type of screw pile, shown in Figure 6-10, is used where good soils are available at a relatively deeper depth. Figure 6-11 shows a wood pole installed inside a screw pile.

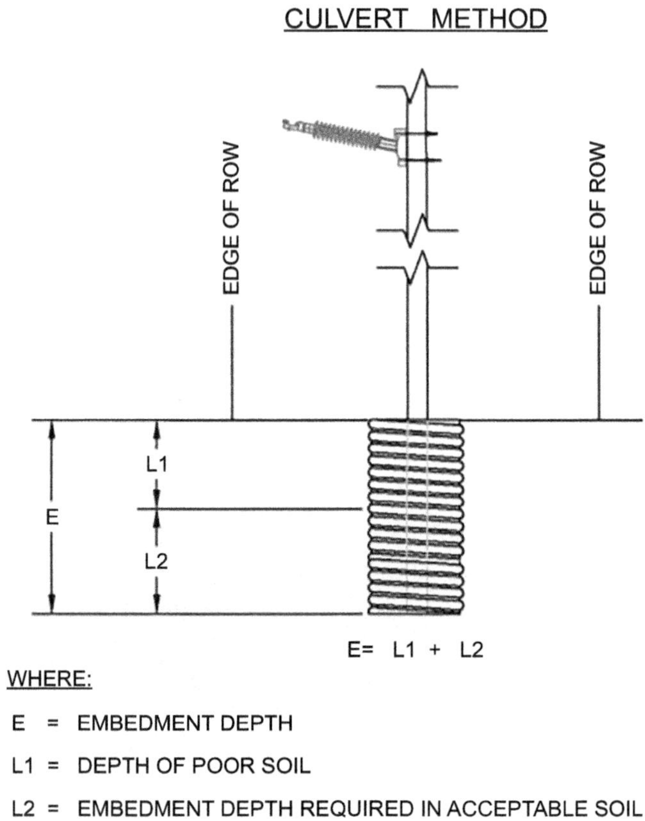

Figure 6-7. Deep set in culvert foundation design.
Source: Great River Energy, reprinted with permission.

6.3.2.5 Driven Piles. Although driven piles can be used for poor soil conditions, their real benefit is in locations where excavation is prohibited because of soil contamination. Typically, the pole would be set on the existing ground and held while the pile sections are driven to the depth determined by the engineer. The pole would then be through-bolted and banded to the exposed section of the driven piles.

6.3.2.6 Bog Shoes. A bog shoe, shown in Figure 6-12, is a platform constructed from timbers or logs to increase the bearing area of the base of the pole. It is an option for poles set in soft, wet conditions (e.g., muck) and is generally only practical for shorter poles in lightly loaded conditions. It may be preferable to use piles or set the structure deeper to reach a stable soil strata rather than use a bog shoe.

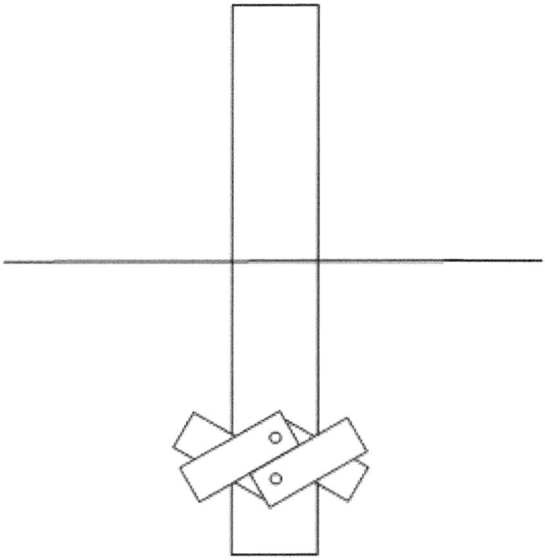

Figure 6-8. Pole key.
Source: Vicki Schneider, reprinted with permission.

Figure 6-9. Screw pile with variable tube diameters.
Source: Kishor Kumar, reprinted with permission.

Figure 6-10. Screw pile with uniform tube diameters.
Source: Kishor Kumar, reprinted with permission.

Figure 6-11. Installed pole inside screw pile.
Source: Kishor Kumar, reprinted with permission.

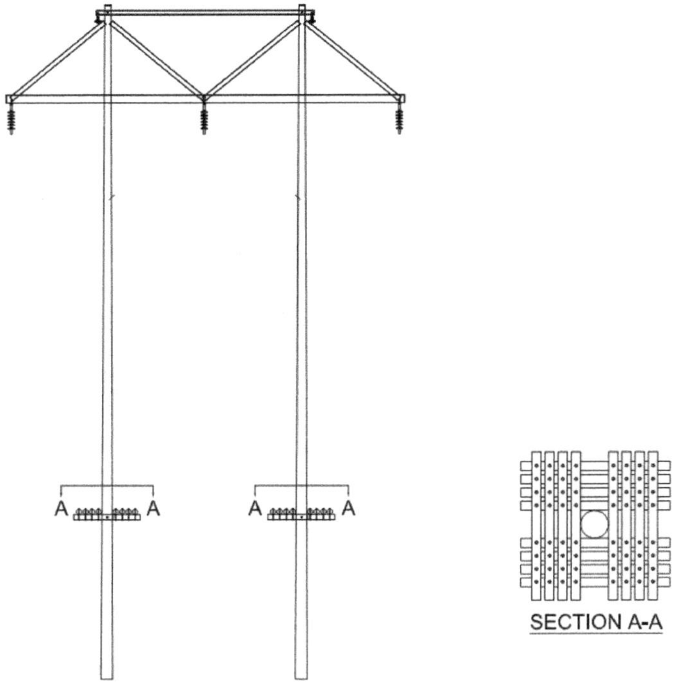

Figure 6-12. Bog shoe.
Source: Vicki Schneider, reprinted with permission.

6.3.2.7 Uplift Brackets. Uplift brackets (see Figure 6-13) are plates attached to the base of the structure used to add surface area to increase the uplift capability of the structure; they can also be used to increase bearing capacity, but values will need to be derated.

6.3.2.8 Bearing Pad. A bearing pad (Figure 6-14) is a layer of gravel compacted at the base of the hole to allow for additional bearing capacity. It is typically required for framed or guyed structures with significant bearing load in areas where the soil does not have sufficient bearing capacity. The depth of gravel required depends on the load, soil type, hole diameter, and gravel type used.

6.4 ANCHORAGE

Continuous loads from wire tensions owing to line angles or dead-end applications cannot be placed on wood poles without guying the load.

Figure 6-13. Uplift brackets.
Source: Yair Berenstein, reprinted with permission.

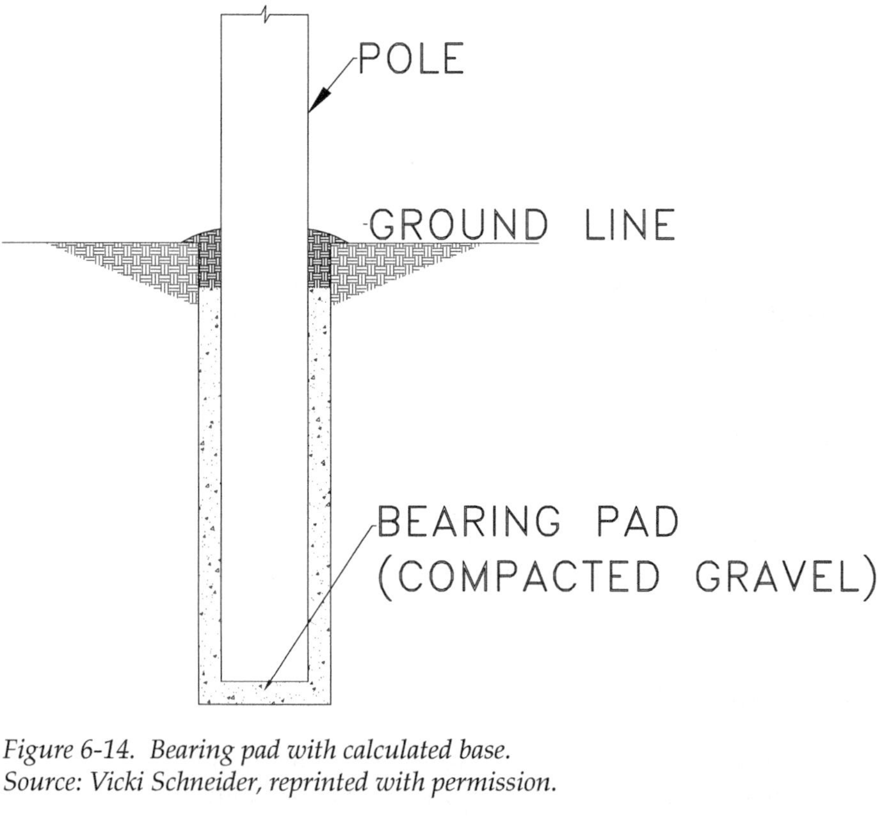

Figure 6-14. Bearing pad with calculated base.
Source: Vicki Schneider, reprinted with permission.

Continuous unguyed loads will cause permanent deformation of the pole. Guy wires direct the loads from the pole to an anchor installed in the ground.

6.4.1 Screw Anchors

A screw anchor (or soil helix anchor) consists of one or more helically deformed plates attached to a central core, as illustrated in Figure 6-15. The anchor is installed by rotating it into the ground, usually with a truck-mounted power auger. Screw anchors can be designed for compression loads as well as for tension.

Screw anchors are made by several manufacturers that typically supply engineering references to aid in the design and selection of the proper screw anchor. These recommendations can be used for normal anchoring situations. For unique situations, the manufacturer should be consulted and any recommendations reviewed and approved by a geotechnical engineer.

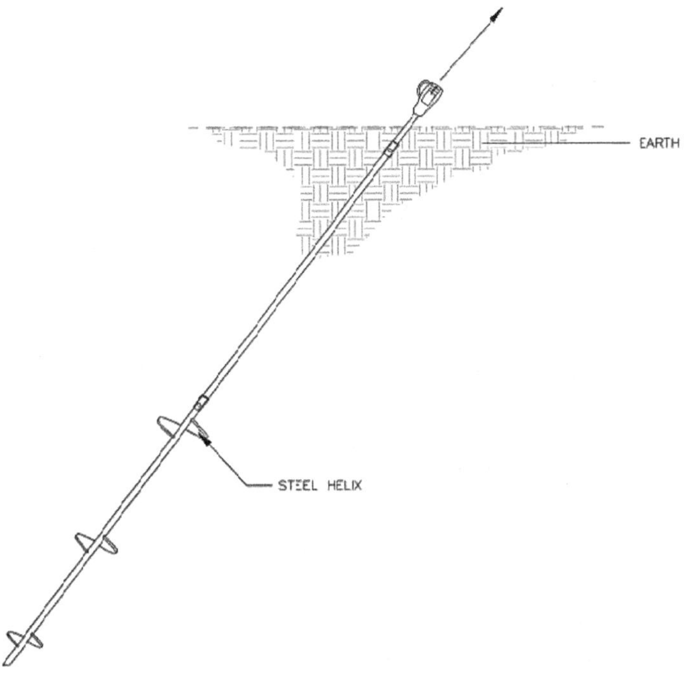

Figure 6-15. Screw anchor.
Source: Great River Energy, reprinted with permission.

6.4.2 Plate and Log Anchors

Plate anchors (Figure 6-16) and log anchors (Figure 6-17) require separate excavations for the anchor and anchor rod. They are placed in excavated holes and backfilled and bear against the undisturbed sides of the hole. The load is applied through the anchor rod to the anchor, causing the anchor to bear on relatively undisturbed earth. Log anchors are sometimes used because of the ease in obtaining material. The capacity of the anchor is limited by the strength of the wood log, plate, or connection between the rod and the anchor.

6.4.3 Spread Anchors

Spread anchors are similar to spread footings; they develop their resistance to uplift by the weight of the anchor plus the weight and strength of the soil above it. Spread anchors are constructed in a pit or trench, as shown in Figure 6-18. Various materials can be used for the plate including steel, reinforced concrete, and precast concrete. This type of anchor is typically used in dry soils containing boulders in which drilled or screw anchors cannot be easily installed.

6.4.4 Drilled Shaft Anchors and Pile Anchors

A reinforced concrete shaft may be used as an anchor (see Figure 6-19). Augered holes of various diameters are filled with cast-in-place concrete. A steel rod or cable is cast into the concrete and transmits the load to the shaft. The movement of the shaft is resisted by its weight plus the friction or adhesion resistance of the surrounding soils. Driven piles may also be used as anchors (Figure 6-20) in a manner similar to the drilled shaft anchor.

6.4.5 Rock Anchors

A rock anchor consists of a steel tendon (either a steel bar, tower leg angle, or steel cable) placed into a hole drilled into the rock. As shown in Figure 6-21, rock anchors work by grouting the void around the steel tendon or by using a mechanical locking mechanism that expands into the surrounding rock at the anchor tip.

Rock anchors are designed by considering the ultimate pullout capacity. This capacity is determined by several interfaces: rock mass, grout-rock bond, grout-steel tendon bond, and steel tendon strength (including connections). Test borings, field inspection of the excavations, and load tests are important to the design of the rock anchors.

A commonly used method for determining the load capacity of an anchor in heavily jointed or weak rock is to assume the rock mass fails along an

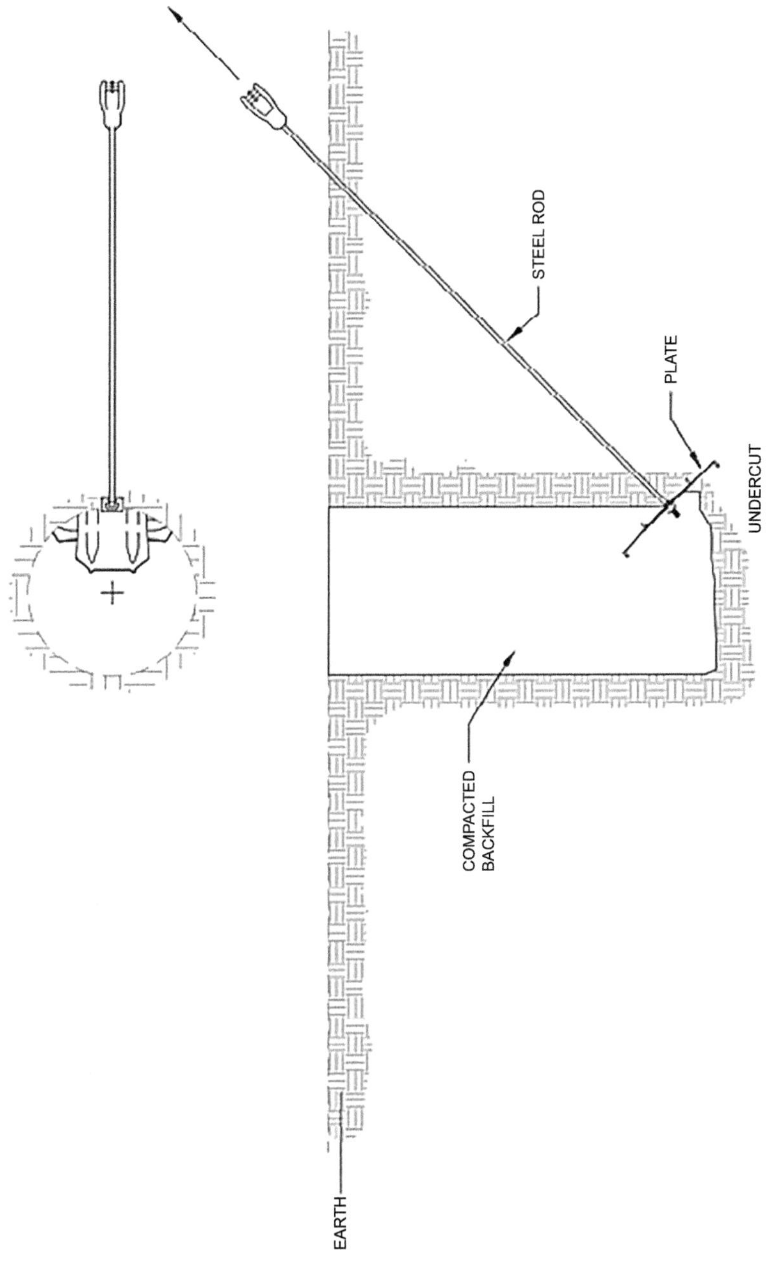

Figure 6-16. Plate anchor.
Source: Great River Energy, reprinted with permission.

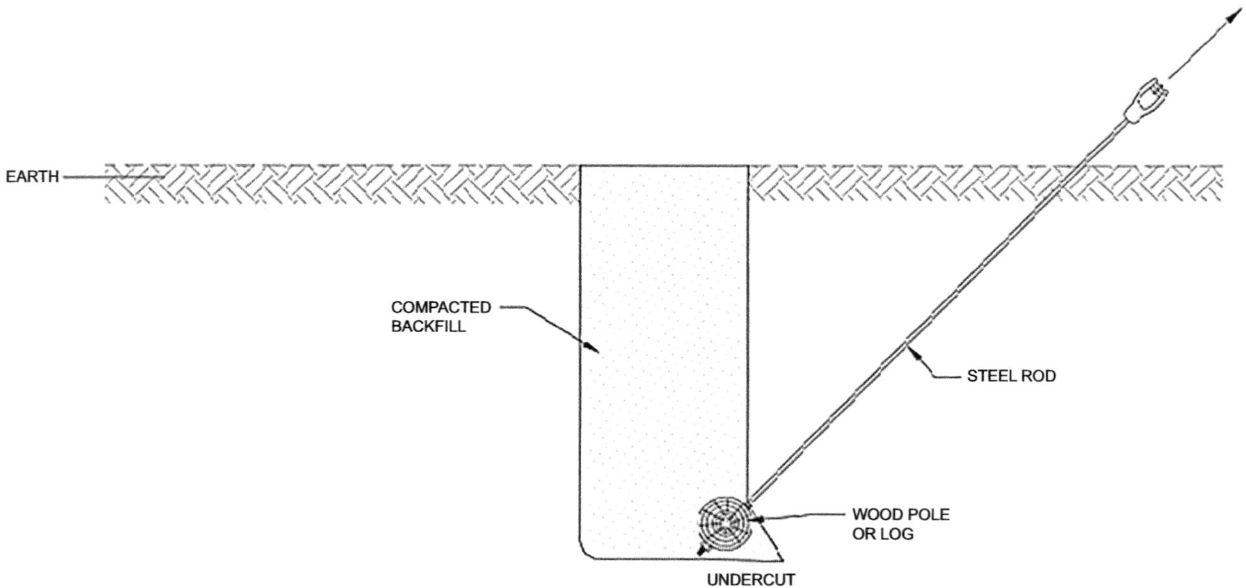

Figure 6-17. Log anchor.
Source: Great River Energy, reprinted with permission.

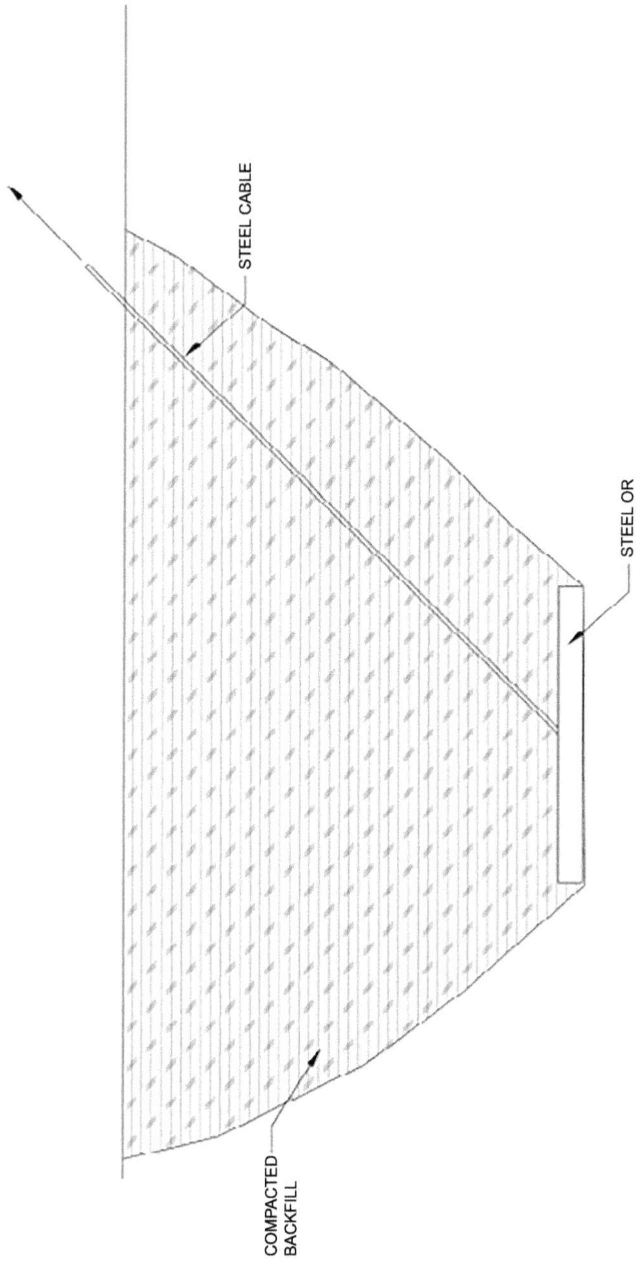

Figure 6-18. Spread anchor.
Source: Great River Energy, reprinted with permission.

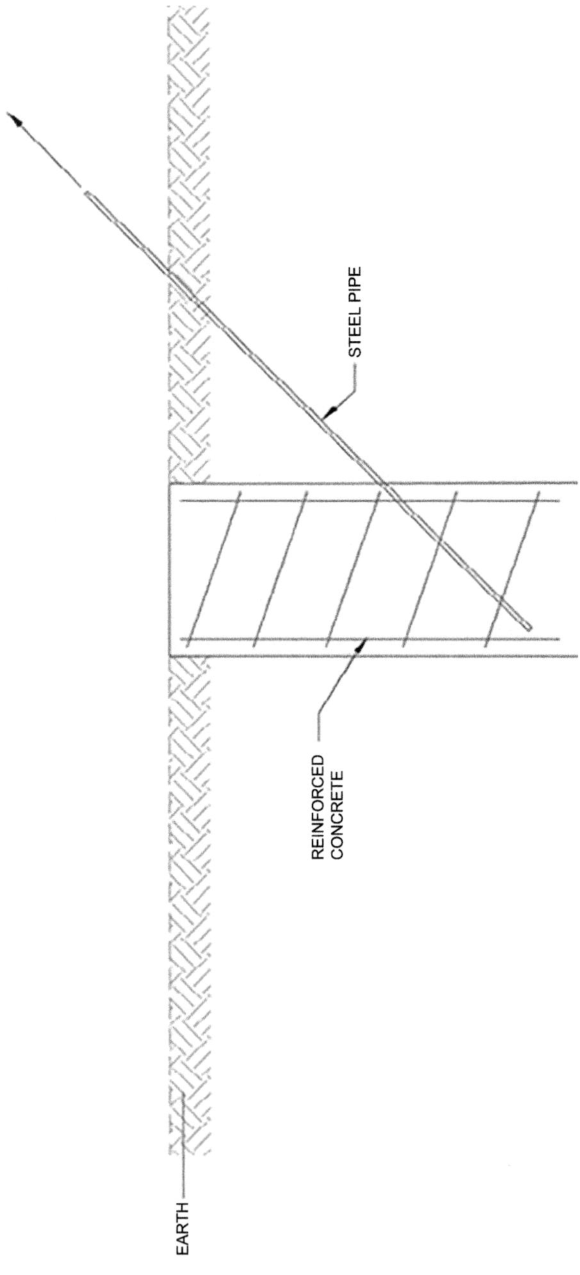

Figure 6-19. Drilled shaft anchor.
Source: Great River Energy, reprinted with permission.

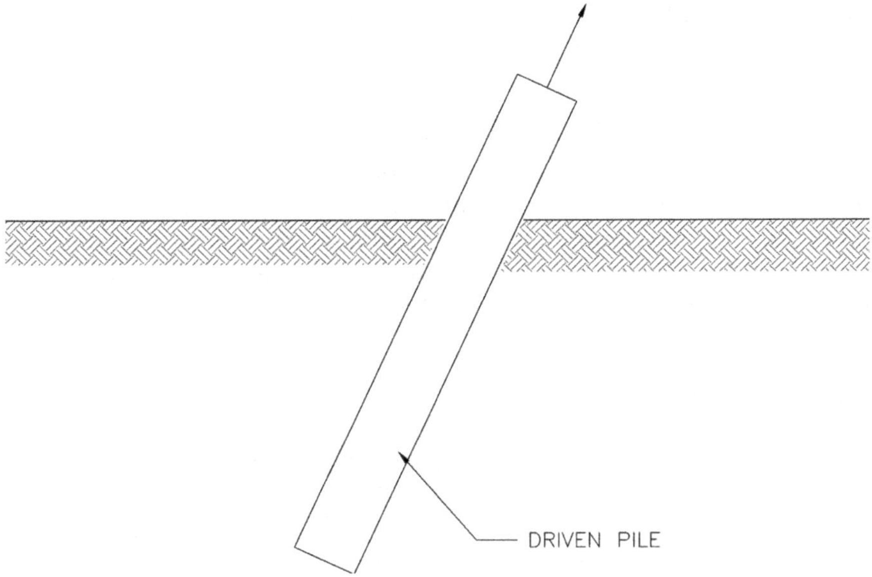

Figure 6-20. Pile anchor.
Source: Great River Energy, reprinted with permission.

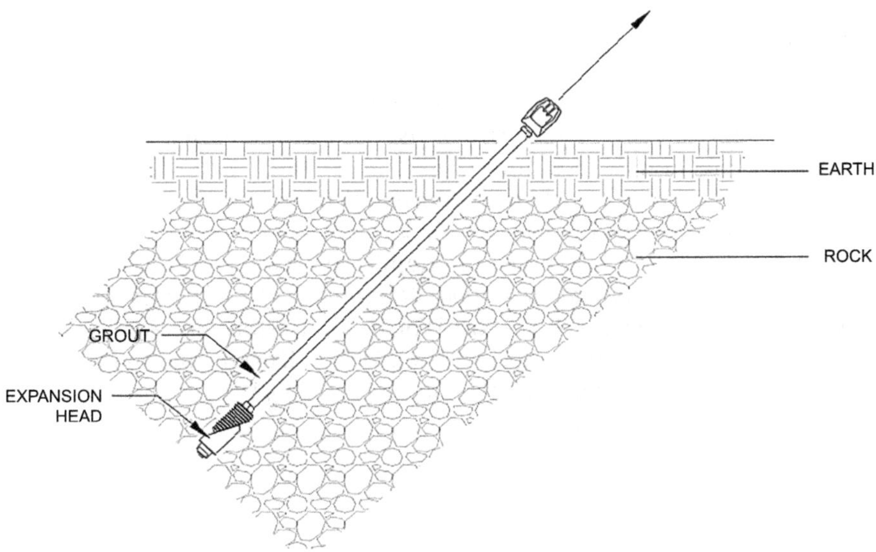

Figure 6-21. Grouted rock anchor.
Source: Great River Energy, reprinted with permission.

assumed shear plane. The load resistance is equal to the weight of the rock plus the shear strength of the rock along the assumed failure plane.

6.4.5 Anchor Installation and Testing

Anchors should be tested to ensure they can accommodate the design load. Soil conditions can vary, greatly affecting the pullout strength of any given anchor. Two common types of testing methods are pull testing and torque testing. Always follow the manufacturer's installation and testing procedures when using anchor systems.

6.5 FOUNDATION SUMMARY

Many variables need to be considered in the design of a wood structure embedment. Among these are lateral loads, vertical loads, soil strength, performance, reliability, deflection, water table, side slope, frost line, and right-of-way considerations. Because of these variables, the standard "rule of thumb" method of 10% of the pole length plus 2 ft (10% + 2) is not recommended for foundation embedment design. Foundations should be designed based on actual loading and project-specific soil conditions.

REFERENCES

ANSI/AWPA (American Wood Protection Association). 2017. ANSI-O5.1, *Wood poles—specifications and dimensions*. Birmingham, AL: American Wood Protection Association.

ASCE. 2010. *Guidelines for electrical transmission line structural loading*, MOP 74. Reston, VA: ASCE.

Lutein, C. M. 2009. "Wood pole bury depths in various soils." *Proc., Electrical Transmission Line and Substation Structures Conf. Technology for the Next Generation*. November 8–12, Fort Worth, TX. Reston, VA: ASCE.

USDA (US Department of Agriculture). 2015. "Design manual for high-voltage transmission lines," *RUS Bull. 1724E-200*. Washington, DC: Government Printing Office.

CHAPTER 7
MANUFACTURING AND QUALITY ASSURANCE

7.1 GENERAL

A natural round wood pole is not an engineered product. The purpose of this chapter is to help the engineer understand components of the manufacturing process and quality assurance for poles to perform as intended. The manufacturing requirements specified in ANSI O5.1 (ANSI/AWPA 2017), Canadian Standards Association CAN/CSA O15 (CSA 2015), and American Wood Protection Association (AWPA) Standard U1, Commodity Specification D: Poles (AWPA 2018), and RUS Bulletin 1728H-702 (USDA 2011) should be closely monitored by the manufacturer to ensure that the quality of the poles meets or exceeds the minimum standards.

The variability of wood poles is quantified by the coefficient of variation (COV). As indicated in Chapter 4, the high COV of wood poles means a greater percentage of the population is able to withstand loads that are higher than the required ultimate strength. This helps wood poles resist extreme loading events better than materials with a low COV. Natural round wood poles, with their inherent qualities and manufactured with effective quality assurance, are a reliable structural material for design for engineers.

7.2 MANUFACTURING PROCESS

The manufacturing process starts by harvesting select timber that is transported to a manufacturing facility. After peeling the bark and additional preparation, poles are sterilized before the pressure treatment process to eliminate internal decay fungi.

Figure 7-1. Forested area where wood poles are harvested.
Source: Bell Lumber and Pole Company, reproduced with permission.

7.2.1 Harvesting Timber in Woodland

The most important step in the pole manufacturing process is the selection of the whitewood. There are many options for obtaining the whitewood. The process can take from one day to eight months, and in some cases even longer. This wide variance of time depends on the species, class, length, and quantity of poles the customer requests. In a stand of trees, approximately 5% to 7% of the selected timber will meet the quality standards of a wood utility pole. The remaining timber is used for dimension lumber, pulp, and other applications (USDA Forest Service 2010).

Trees in the forest (as shown in Figure 7-1) are selected for utility poles by a forester and are marked for the anticipated class and length. The basal area is the space that allows other trees to grow in the optimal spacing after a tree is marked for harvest. The forester follows accepted practices by selecting trees within a basal area. The circumference at 6 ft from the butt is measured to determine the expected class of the pole. The height is determined by either laser measurement or visual approximation by the forester.

The tree is then clearly marked for the harvester/logger to see and is written into a logbook, with its circumference measurement, to keep an accurate account. Figures 7-2 and 7-3 show that the soon-to-be pole is cut and de-limbed in the forest by machine. It is then put on a logging trailer and trucked to the peeler for inspection and verification of the class and length, before final manufacturing.

Figure 7-2 (left) and Figure 7-3 (right). Pole harvesting from a basal area.
Source: Bell Lumber and Pole Company, reproduced with permission.

Figure 7-4. Peeling process.
Source: Bell Lumber and Pole Company, reproduced with permission.

7.2.2 Peeling

Peeling or debarking the pole is performed by a machine-operated rotary blade and a conveyer system as shown in Figure 7-4. This operation is typically controlled by a skilled worker to maintain proper depth of the debarking process.

The bark is then reclaimed and used for landcaping mulch, animal bedding, or hog fuel for cogeneration plants. The poles are sorted and inspected per the specifications (ANSI, AWPA, CSA, or RUS) as explained in Section 7.1. At this point the poles are termed to be "in the white" as shown in Figure 7-5.

7.2.3 Preparation for Treatment

The next step in manufacturing poles is the preparation before conditioning and treatment. This is when the final class and length are established, marking/branding takes place, the groundline zone for Douglas fir and Western red cedar poles is defined and prepared, the predrilling of framing holes is performed, and final whitewood inspection is completed.

Drilling the framing holes before treatment helps to ensure effective treatment and proper hole drilling and may save time in the field for installation.

Figure 7-5. Wood poles after peeling.
Source: Bell Lumber and Pole Company, reproduced with permission.

One caveat is that the structure framing drawing details given to the manufacturer must be clear, correct, and easy to implement. Tolerances should be defined by the engineer for the manufacturer to ensure that the holes are properly located and oriented on the pole.

The predrilled holes are bored by hand, machine, or precision drilling. If the tolerances are tight, a precision-drilled hole is the best. As tolerances get larger, machine- and hand-drilled methods are adequate. It is important to note that precision drilling takes more time and affects the rate of production for the manufacturer, which in turn also adds to the overall cost.

Poles made of Douglas fir and Western red cedar need additional preparation in the groundline zone to improve the penetration of the preservative in areas where future decay is most likely. The engineer must therefore calculate the required embedment depth prior to placing the pole order and specify the length of the zone above and below the expected setting depth for additional pretreatment processes. The manufacturer will label the poles according to the results of the final manufactured product. If no instructions are given about the location of the groundline zone or the required embedment, the manufacturer will default to a groundline location of 10% of the length of the pole plus 2 ft.

Through-boring, radial boring, and incising are three processes commonly used to prepare thin sapwood species for treatment. The heartwood does not accept preservative during the treatment process, so these preparation methods enhance the penetration. Engineers must consider in their design whether these processes have any effect on strength.

Douglas-fir poles are most commonly prepared for treatment with through-boring and radial boring, although incising can be performed. Figure 7-6 shows a typical through-boring pattern, and Figure 7-7 shows a pole after through-boring is completed. Figures 7-8 and 7-9 show a typical pattern for radial boring and poles after radial boring is completed.

It is more common for Western red cedar poles to be prepared for treatment in the groundline zone with incising. This process perforates the pole with small knives or pegs to open avenues of penetration to ensure uniform and adequate preservative distribution. The depth of the incising ranges from 5/8 in. to 1 in. and typically extends 1 to 2 ft aboveground and 2 to 4 ft belowground. Figures 7-10 and 7-11 show a pole during and after incising.

7.2.4 Seasoning and Conditioning

Poles that are termed "in the white" are also considered to be in a green state, indicating a high moisture level in the pole. Air seasoning is recommended for Western red cedar so that the poles can dry from the green state. This is accomplished after the poles are inventoried and stored for

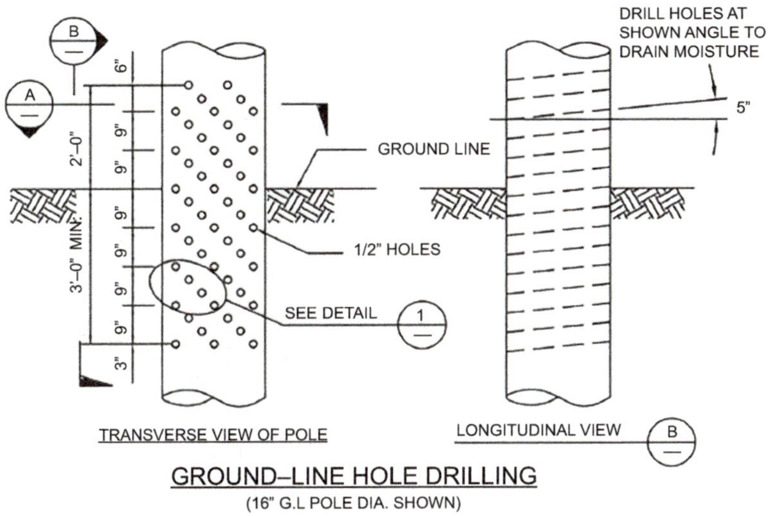

Figure 7-6. Drawing of through-boring drilling pattern in the groundline zone. Source: Bell Lumber and Pole Company, reproduced with permission.

Figure 7-7. Picture after through-boring drilling is completed. Source: Bell Lumber and Pole Company, reproduced with permission.

drying. In Figure 7-12, the natural lengthwise separation of wood that usually extends across the rings of annual growth, commonly referred to as splits or checks, will open wider as moisture leaves the pole and exposes the maximum amount of internal wood to receive preservative treatment.

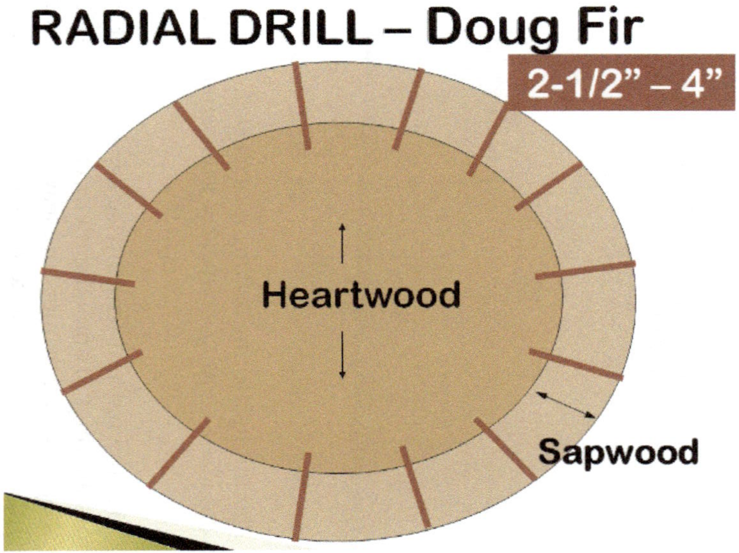

Figure 7-8. Cross-section of radial drill pattern.
Source: Bell Lumber and Pole Company, reproduced with permission.

Figure 7-9. Radial drill patterns in the groundline zone.
Source: Bell Lumber and Pole Company, reproduced with permission.

Figure 7-10. Pole during incising.
Source: Bell Lumber and Pole Company, reproduced with permission.

Figure 7-11. Pole after incising.
Source: Bell Lumber and Pole Company, reproduced with permission.

Figure 7-12. Freshly kiln-dried and treated poles with a normal amount of checking. Source: Nelson Bingel III, reproduced with permission.

This method of drying is listed in Table 1 of ANSI O5.1 as Treatment Group A. The four treatment groups listed in the ANSI standard are as follows:

- Group A: Air Seasoning—Poles are stacked and allowed to dry naturally, as shown in Figure 7-13.
- Group B: Boulton Drying—Water is extracted by heating the material in the preservative while placing the treating retort under vacuum and condensing and quantifying the moisture removed.
- Group C: Steam Conditioning—Water is extracted using steam to heat the poles to a specific temperature for a specific duration, depending on the dimensions of the pole.
- Group D: Kiln Drying—The temperature is increased gradually to a specified value. The rate of the temperature increase and the maximum temperature to which the poles are exposed depends on the wood species.

ANSI O5.1 Table 1 also indicates which treatment group should be used for each wood species. Southern pine is listed in both Treatment Group C and Group D, with kiln drying most commonly used. Douglas fir is listed in Treatment Group B and Group D, with Boulton drying most commonly

Figure 7-13. Air seasoning using "chimney stacking."
Source: Bell Lumber and Pole Company, reproduced with permission.

used. Western red cedar is listed in both Treatment Group A and Group D, with air seasoning most commonly used.

Sterilization is the last step before applying preservative treatment. Western red cedar is the only species that is not required to be sterilized because of its high durability and natural resistance to decay. The sterilization process is required for all other species. When manufacturers are required to sterilize the wood, they heat the wood fibers to at least 150°F at the center of the entire pole for at least one hour. This process kills any microorganisms that will contribute to decay.

7.2.5 Treatment Process

The original preservative treatment of wood poles helps to protect against decay and insect infestation, which extends the pole's useful life. Preservatives for wood poles are categorized as either oil-borne or water-borne preservatives. Oil and water are the carriers of the preservatives into the wood cells by means of pressure to penetrate deep into the wood cells. The most common treatments are the following:

- Oil borne—pentachlorophenol (PCP), creosote (CR), copper naphthenate (CuN), copper azole (CA), and
- Water borne—chromated copper arsenate (CCA), alkaline copper quat (ACQ), and ammoniacal copper zinc arsenate (ACZA).

The basic sequence by which the preservative is applied in a pressurized cylinder (shown in Figure 7-14) is typically as follows:

- Initial vacuum or initial air pressure,
- Fill cylinder,

Figure 7-14. Treatment cylinder.
Source: Bell Lumber and Pole Company, reproduced with permission.

- Pressure injection,
- Expansion bath (Western species only), and
- Final vacuum.

After the poles are removed from the treating cylinder, they are put in holding areas to be inspected for both penetration and retention of preservative. In the case of water-borne treatment, such as CCA, a certain amount of time for the poles to "fixate" is required. This is a complex chemical process because of the time and temperature needed for the chemical to bond to the wood fibers, but is necessary before the poles can be transported.

7.3 QUALITY ASSURANCE/QUALITY CONTROL

Any good quality-control program at a wood utility pole supplier encompasses two basic disciplines: physical material quality and preservative treating results. Each of these disciplines are species- and preservative-dependent and are recognized as minimum requirements. Specific aspects of these disciplines are explained in the following sections.

7.3.1 Material Quality

Each wood species presents its own inherent quality characteristics (e.g., durability, fiber strength, etc.) that affect the handling, treatment, and performance requirements of the utility pole. These characteristics are taken into consideration when establishing national standards and customer specifications depending on pole utilization. For example, a utility company may tighten appearance requirements if the pole(s) are going to be placed in an upscale neighborhood or public park; these requirements always take precedence over the industry-acceptable (minimum) standards. The most widely accepted wood pole standard for material quality is ANSI O5.1 (2017).

7.3.2 Treating Results

The natural characteristics of the wood will determine which preservative(s) can be used in treatment, what penetration depths are required, and how much chemical must remain post-treatment (known as "retention") to provide adequate protection for the application. The most widely accepted wood treatment is Standard AWPA T1, which is reviewed and updated annually as necessary. To summarize:

- Preservative—Each species has approved preservative(s) that have proven efficacy. These preservatives are combined with either a water- or oil-borne carrier and treated at varying temperatures and/or pressures to ensure the wood cells can be penetrated.
- Penetration—Each species has required preservative penetration depths to ensure adequate protection for the pole once in service. Penetration can be measured either visually or by application of a chemical reagent after an increment core is taken perpendicular to the axis of the pole. Since the wood pole is a natural product with varying growth rates, it is difficult to control penetration depth from pole to pole. Pole buyers can expect to have a greater preservative penetration than required, which translates to extra protection against decay and insects.
- Retention—A chemical analysis ("assay") is conducted on specific segments of the increment core (based on species, preservative, and application) to isolate the active chemical(s) that make up the preservative. The analytical method varies based on the active ingredient(s) in the preservative formulation.

REFERENCES

ANSI/AWPA (American Wood Protection Association). 2017. ANSI-O5.1, *Wood poles—specifications and dimensions*. Birmingham, AL: American Wood Protection Association.

AWPA. 2018. T1-18, *Use category system: Processing and treatment standard*. Birmingham, AL: American Wood Protection Association.

AWPA. 2018. U1-18, *Use category system: User specification for treated wood*. Birmingham, AL: American Wood Protection Association.

Canadian Standards Association (CSA). 2015. Standard CAN/CSA-O15, *Wood utility poles and reinforcing stubs*. Mississauga, ON: CSA Group.

USDA (US Department of Agriculture) Forest Service. 2010. *Wood handbook*. Madison, WI: Forest Products Laboratory, chapter 8, 8-1–8-28.

USDA. 2011. "RUS specification for quality control and inspection of timber projects." *RUS Bull. 1728H-702*. Washington, DC: Government Printing Office.

CHAPTER 8
ASSEMBLY AND ERECTION

8.1 GENERAL

This chapter discusses the details involved in handling, transporting, assembling, and erecting wood pole structures. Because there are several manufacturers and a variety of species of wood poles, there may be variations in the handling, assembly, and erection recommendations, thus no one set standard can be established. Rather, the discussion in this chapter is generalized in nature and serves to make the engineer aware of the factors that must be considered. When purchasing poles from a specific manufacturer, it is highly recommended that the engineer follow the manufacturer's guidelines for handling, assembling, and erecting those poles.

8.2 STORAGE

Wood poles are durable and can be stored in practically any location ranging from covered warehouses to open-air pole yards.

Wood poles are usually delivered from the manufacturer with cribbing between the poles to facilitate unloading of the poles. The poles can be stored with or without cribbing. Treated timber sections should be used to keep the poles elevated above the ground both to prevent dirt or other debris from contaminating the poles and to provide an access location to lift the poles by forklift or lifting straps. For poles that are stored horizontally, supports shall be spaced accordingly to prevent sagging or bowing. Wood poles should be blocked to prevent any rolling.

Wood poles that are stored for long periods of time should be rotated. Rotating minimizes the development of a permanent bow in the poles and

Figure 8-1. Wood pole handling with a clamp.
Source: Bell Lumber and Pole Company, reproduced with permission.

ensures that any oil-borne preservatives are evenly distributed throughout the cross-section of the poles. Oil-borne preservatives can migrate to the lowest points on the pole because of gravity. It is recommended that the poles be rotated 180 degrees every six months. The engineer may consult with the manufacturer on the amount of the rotation and the period between rotations for their specific product.

8.3 HANDLING

Because wood poles are very durable, they can take a moderate amount of abuse without being damaged. They can be handled in a variety of ways.

Wood poles can be loaded, moved, and unloaded using a forklift, knuckle boom, or clamp (Figure 8-1). When using a forklift, the pole can be placed directly on the forklift tines. The pole should be positioned with the tines perpendicular to the longitudinal axis of the pole. The pole should be balanced such that its center of gravity shall be between the forklift tines. In the case where wood poles are strapped in a bundle, the center of gravity of the bundle shall be placed between the forklift tines.

Wood poles can also be handled with a boom or crane using nylon slings attached to the pole(s). When using nylon slings to lift a pole, a two-point pickup system is safest, and placement at the center of gravity is not as critical as when using a single nylon sling. If using a single nylon sling, a lineworker must always maintain contact with one end of the pole to ensure that balance is maintained while moving. Nylon slings should always be used in lieu of chains, cables, or other metal hardware when lifting wood poles to avoid damage to the pole's surface. Pole dollies may be used to transport poles as needed. Prior to handling any poles, it is important to verify that the equipment to be used is properly rated to safely carry the anticipated loads.

When handling poles, care should be taken to minimize excessive damage to the poles. Minor scratches and scrapes to the surface of the poles are acceptable. However, any damage that could affect the mechanical properties of the pole, such as gouges or cracks, needs to be inspected and evaluated. The engineer should reference ANSI O5.1 Section 8.3 (ANSI/AWPA 2017) for guidance in these evaluations as well as consult with the manufacturer. Any damage to the surface of the pole that exposes untreated wood should be remedially treated with a preservative to ensure the long-term protection of the pole.

8.4 TRANSPORTATION

Because wood poles are single-piece poles, trucks are needed for transporting the poles. The utility must ensure that the appropriate transportation method is used to safely and legally transport the poles from the material yard to the job site. The transportation of wood poles can range from being relatively simple for the shorter distribution sizes to more complex for poles up to 125 ft. For shorter distribution-size poles, a utility's service truck may be used. For sizes larger than distribution, flatbed trucks, semitrailer trucks, and adjustable pole trailers must be used (Figure 8-2). In the case of the longer poles, road permits and escort vehicles may be required. Caution flags need to be used at the ends of the pole. The manufacturer and the local transportation department should be consulted before transporting long poles to ensure that all proper requirements are followed and permits are obtained.

Care must be taken when planning the transportation route. Depending on the length of poles being transported, road conditions and turning radiuses need to be reviewed to ensure that the truck carrying the poles can make the turns along the route without having a negative impact along the route.

Wood poles shall be secured to the truck during transportation. Cribbing should be used to properly space and minimize shifting and damage to the poles during transit. Cribbing should be evenly spaced and placed on the truck bed with the poles being placed on the cribbing. Chains or nylon

Figure 8-2. Wood pole transportation.
Source: Bell Lumber and Pole Company, reproduced with permission.

straps should be used to secure the poles to the truck. When using chains, care should be taken to minimize any potential damage to the poles. For short distances, pole dollies and other pole-handling vehicles can be used.

8.5 POLE CUTTING

Wood poles can easily be cut in the field. The two primary reasons for cutting a pole are (1) to achieve a pole of a specific length or (2) to obtain a stronger pole for a given pole length. Wood poles come in lengths of 5 ft increments. Should a pole length be required that is outside the 5 ft increment, a longer pole of the same class can be cut to meet the required length. If a given pole application requires a pole that needs to be stronger than the available poles for a given length, the top of a longer pole with the same or greater class can be cut to the required length to provide greater strength. Before cutting a pole under these conditions, the engineer must be consulted to verify that the cut pole will meet the strength requirements of the given load condition. The engineer should consider both the pole strength and the embedment depth to verify the cut pole will meet all requirements.

Wood poles can be cut with a chain saw or other tool that will create a smooth surface and not cause damage to the pole. Poles should only be cut at the tip, as the groundline section of the pole may have additional preservative treatment. After the cut is made, the exposed cross-section must be treated with a wood preservative before the pole is installed. An angle or roof cut could be made to ensure that rainwater will not pond at the top of the pole. A pole cap can also be added to prevent water from intruding into the pole as shown in Figure 8-3.

Figure 8-3. Pole cap.
Source: Osmose Utilities Services, Inc., reproduced with permission.

A star lock can also be installed at the pole top to keep the cross-section intact and minimize the propagation of cracks along the length of the pole.

8.6 ASSEMBLY

Since wood poles are single-piece poles, they have the advantage of needing minimal assembly during construction. For single-pole structures, assembly involves attaching the insulators, crossarms, davit arms, grounding system, equipment, and other wire attachment hardware. In some cases, all these attachments are necessary, and in others, only a few of them are needed.

A framed structure requires much more effort to assemble. These structures can either be assembled in the air or assembled on the ground and erected as a whole. Where possible, assembly on the ground is preferred because it allows for better quality control for all connections. Should a frame need to be assembled in the air, the main vertical members need to be plumbed and aligned before the foundations are set. Once the main vertical members are set, crossarms and braces can be installed. Crossarms can be predrilled or drilled in the air. When erecting a frame in the air, it is preferred that the bolt holes in the crossarms be drilled in the air. This method will ensure a good connection without creating undue stress. The installation of a predrilled crossarm may be challenging if the vertical members are not plumb or in proper alignment. When assembling a frame, wood-to-wood contact between frame members is not recommended. Appropriate hardware should be used at all wood-to-wood connection points.

While most standard cleated and noncleated attachment hardware can be used on wood poles, wherever possible cleated hardware should be used to ensure a firm connection to the pole.

8.7 CONNECTIONS

When connecting hardware to the pole, two types of connections can be used: through bolts and pole bands. Through bolts run completely through the pole cross-section and provide a strong, positive connection. Pole bands are metal bands or straps that are cinched and clamped around the outer circumference of the pole. During installation, the bands are tightened until firm contact is established with the surface of the pole. When using pole bands, periodic inspections should be performed to ensure that a good connection is maintained over the life of the pole because natural wood shrinkage could cause the bands to loosen. When installing through bolts or pole bands, do not overtighten the bolts or bands because it may cause localized crushing of the wood fibers that are in direct contact with the bolt head or band.

Staples, nails, and lag bolts should be used only to secure ground wire and non-load-bearing moldings and markers to the pole.

8.8 FIELD DRILLING

Wood poles are usually predrilled by the manufacturer, but they can also be easily drilled in the field using standard equipment and the appropriate wood drill bits. If a substantial number of holes are to be drilled, an engineer or the manufacturer should be consulted to ensure that the strength of the pole is not compromised.

The manufacturer's recommendations shall be followed in the field drilling process. In general, the minimum edge-to-edge distance between holes shall be six times the diameter of the holes. Any holes that are not used should be plugged to prevent moisture or insects from entering the pole. The hole should first be treated with a preservative, then plugged with either a plastic plug or a treated wood plug. An engineer should be consulted to review any hole clusters to ensure the continued strength of the pole in the area of the cluster.

8.9 POLE GAINING

Gaining is the process in which a flat surface is cut into the side of the round wood pole. This flat surface allows for better contact between the pole and a flat-sided crossarm or between the pole and flat-sided pole hardware. Any required gains should be cut at the manufacturer's facility prior to treatment. Gains shall not be cut in the field. Should it be determined that a gain is required in the field, the gain should be achieved by the use of appropriate gain hardware.

8.10 ERECTION

Conventional equipment and practices can be used as appropriate for the terrain and site conditions at the pole location. Single wood pole structures can be erected by crane using a single pickup point as recommended by the manufacturer. Nylon slings are commonly used to secure the pole to the lifting cable to avoid damaging the pole surface.

As the pole is being lifted, tag lines can be used to guide it to the foundation. Once the pole is in place, its vertical plumb or rake should be checked before securing the pole in the foundation.

Wood poles can also be erected by gin pole methods whereby either a temporary pole or an existing pole can be rigged with pulleys and hardware to lift and erect the new wood pole.

In remote, inaccessible areas or in areas with challenging terrain conditions, wood poles may be carried and erected using helicopter techniques. Care must be taken to ensure that the weight of the pole is within the safe operating limits of the helicopter.

Once the pole has been set, temporary guys may be required to hold the pole in place while the pole is framed and the conductors are strung and tensioned. Depending on the situation, if a pole is not restrained during the stringing process, the deflection of the pole could affect the final tension and sags of the conductor. This is particularly important if the wood pole or structure is a dead-end or stringing dead-end structure. Once all construction activities have been completed, the temporary guys may be removed.

8.11 CLIMBING

Wood poles can easily be climbed. Safety is the number one priority when climbing poles. If climbing is required, fall-protection devices shall be used. Before climbing a pole, the lineworker should be familiar with and trained in the company's safety rules and procedures. In addition, the appropriate Occupational Safety and Health Administration (OSHA) requirements should be followed.

8.12 DISPOSAL

Used wood poles can be disposed of in many different manners. Wood poles can be cut into manageable lengths and placed in disposal bins. Final disposition of the pole sections depends on the wood pole type and wood preservative treatment. Company standards and local governmental environmental regulations and requirements need to be consulted to determine the best practice to dispose of the wood pole sections.

If utility company and local regulations allow, wood poles can also be given away to local public interests for secondary uses. During the transfer process, Safety Data Sheets (SDS) on the wood pole preservative shall be provided to the customer. The company's legal department should also be consulted to ensure that liabilities are released during the transfer process.

8.13 MISCELLANEOUS CONSIDERATIONS

Because wood poles are a natural product, they are susceptible to damage from natural conditions. The discussion in this section is high level and covers a few of the most common damaging conditions, with recommendations to protect the poles against them. Should a company have problems with any of these conditions, third-party organizations with detailed knowledge of these specific conditions should be consulted to develop any remedies or protective measures.

8.13.1 Woodpeckers

Woodpeckers can cause significant damage to a wood pole (Harness and Walters 2005), which can reduce the pole's structural integrity and may require restoration or replacement. Wire mesh can be placed around the pole in the areas prone to woodpeckers. The mesh offers a measure of protection to critical areas of the pole and, depending on species of woodpecker, can prevent severe damage from nesting cavities.

Another option for areas that are prone to woodpeckers is to consider alternate pole material types such as concrete, steel, and fiber-reinforced polymers (FRP). Although these types of poles can be costlier than wood poles, the overall life-cycle cost may be lower than the cost of having to change out the wood poles periodically because of woodpecker damage.

8.13.2 Termites

Termites can also cause significant damage to a wood pole. The extent of the damage depends on the type of termite, the wood species, and the extent to which the wood preservative penetrates into the pole. There are two basic defenses against termites: barrier methods and wood preservative methods.

Barrier methods are effective against subterranean termites. They protect a wood pole by creating a physical barrier between the termites and the pole. Examples include

- A termite barrier backfill in which the voids between the particles of the backfill are too small for a termite to crawl through,
- Placing the embedded portion of the wood pole in an impermeable vinyl "sock," and
- Placing the embedded portion of the wood pole in a steel mesh "sock" whose openings are too small for a termite to crawl through.

The advantage of using the steel mesh is that while it keeps termites away from the pole, any water that may pass through the pole can be allowed to filter back into the ground.

Wood preservative methods protect the pole by applying the wood preservatives into the heartwood of the pole. This is done by through-boring the portions of a pole that are susceptible to termite intrusion. In this method, small holes are drilled through the pole typically in sections around the groundline. These holes are drilled before pressure treatment so that during the pressure treatment process, the wood preservatives are able to enter the heartwood area and saturate the inside of the pole. Since there is no one uniform pattern for through boring, it is recommended that the utility consult with the wood pole supplier to determine the best through-boring pattern for the particular site conditions.

8.13.3 Fire

Wood poles are susceptible to damage from brush or forest fires. There are many products that can protect wood poles from fire. They are categorized as either coatings, wraps, or barriers (Marquardt 2013). A coating is a product that can be coated or painted onto a pole to provide fire resistance. A wrap is a product that is physically wrapped around a pole, and a barrier is a product that shields a pole from a fire. In the case where a pole is in an area of brush, the coating or wrap should be applied from the groundline to a point approximately 2 ft above the brush height.

When selecting a fire-retardant product, the utility must ensure that the product does not interfere with its use and work methods. For instance, should the pole be located in an inaccessible area and lineworkers need to be able to climb it, the fire-retardant product selected should allow climbing without significantly reducing the protective characteristics of the product. It is recommended that the utilities consult with the wood pole supplier in properly selecting the optimum fire-protection method for their particular site conditions.

REFERENCES

ANSI/AWPA (American Wood Protection Association). 2017. ANSI-O5.1, *Wood poles—specifications and dimensions*. Birmingham, AL: American Wood Protection Association.

Harness, R. E., and E. L. Walters. 2005. "Knock on wood: Woodpeckers and utility pole damage." *IEEE Indus. Appl. Mag.* 11 (2), 68–73.

Marquardt, R. 2013. "Protecting wood utility poles from wildfire." Accessed November 16, 2018. https://www.utilityproducts.com/articles/print/volume-17/issue-3/product-focus/line-construction—maintenance/protecting-wood-utility-poles-from-wildfire.html.

CHAPTER 9

INSPECTION, ASSESSMENT, MAINTENANCE, AND REPAIR

9.1 GENERAL

A major focus for the owners of overhead wood pole lines is optimizing the service life of existing facilities while ensuring safety and structural reliability and resiliency. An effective inspection, maintenance, and repair program is key to achieving this objective. Because of variation in pole species, preservative treatment, environmental conditions, and structure types, pole owners develop customized, cost-effective strategies for inspecting and maintaining their wood pole assets. This chapter provides an overview of relevant issues and processes.

9.2 REGULATORY INSPECTION REQUIREMENTS

Section 214 of the National Electrical Safety Code (IEEE 2017) and many state regulatory bodies require pole owners to perform regular inspection and maintenance of their structures. The cost of inspection and maintenance is a small percentage of pole replacement costs and can markedly prolong the useful life of the system, thereby reducing life-cycle costs. Most important, it also maintains the structural reliability and resiliency of the system while improving safety.

There is no single inspection frequency that works for every overhead line because the processes of deterioration involve a multitude of factors. Lines in cool, dry climates tend to have a lower risk of degradation than lines in hot, humid locations. Other factors to consider when establishing an inspection cycle include wood species and initial preservative treatment, pole age, previous inspection activity, and prior use of remedial treatments.

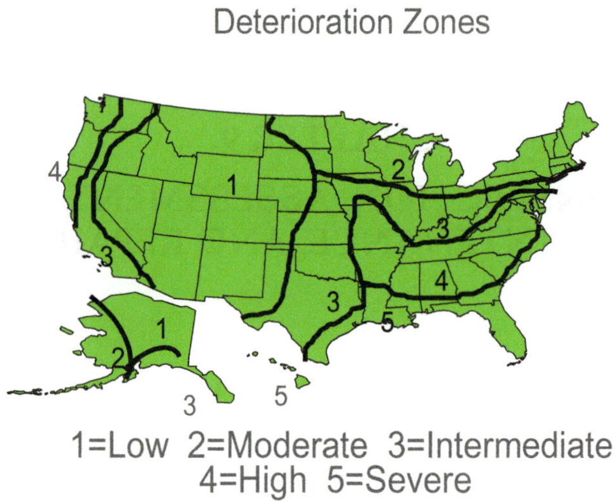

Figure 9-1. Decay hazard zones for the United States.
Source: American Wood Protection Association (AWPA).

The US Rural Utilities Services (RUS) Document 1730B-121 (USDA 2013) recommends inspection cycle lengths ranging from 8 to 12 years based on decay hazard zones. Figure 9-1 is from the American Wood Protection Association standard depicting the decay hazard zones for the United States. Increasing levels of decay hazard call for shorter inspection and remedial treatment cycles. These recommendations are focused on controlling and preventing decay in the groundline zone and do not consider some of the other agents of deterioration, such as woodpecker damage. The American Wood Protection Association Standard M13 (AWPA 2015) can be referenced for additional information on wood pole inspection and maintenance practices.

9.3 AGENTS OF DETERIORATION AND DAMAGE

Wood is a biological material that is susceptible to degradation from a variety of sources, with decay in the groundline zone being predominant. Several different insect species also attack wood poles, with carpenter ants and termites causing the most damage. Insect damage is also most prevalent in the groundline zone of a pole.

Mechanical damage usually occurs above the groundline and can be caused by animals rubbing against the wood or clawing or gnawing the pole or by contact with vehicles, farm machinery, or other heavy equipment. Low-level grass fires that char the outer shell of a pole are another source of pole deterioration. Mechanical and fire damage are generally near the groundline and can be detected during visual inspection.

Pole top damage may occur because of splitting, decay, lightning strikes, and woodpeckers. Evaluation of the extent of these types of damage may require a climbing or aerial inspection, as described in later sections of this chapter.

9.3.1 Groundline Decay

The original preservative treatment of a pole helps prevent decay from establishing in the groundline zone. However, after years of in-service exposure to the environment and degradation mechanisms, the levels of the decay-resisting ingredients may drop below threshold values. Wood without effective preservative treatment becomes a food source for decay fungi when adequate amounts of moisture and oxygen are present along with the necessary temperatures.

Most of the moisture in the groundline zone of a pole results from water near the butt of the pole that moves upward through capillary action within the wood fiber structure, as shown in Figure 9-2. Once the moisture migrates above groundline it tends to dissipate to the atmosphere, leaving only a cone of wetness aboveground. Above this cone of wetness there often is not enough moisture within the pole for decay to establish, except in very wet climates.

In the case of oxygen, there is a depth underground, below which the soil does not typically contain adequate oxygen to support decay. In most soil types the depth is 18 in., but there are combinations of soil types and pole

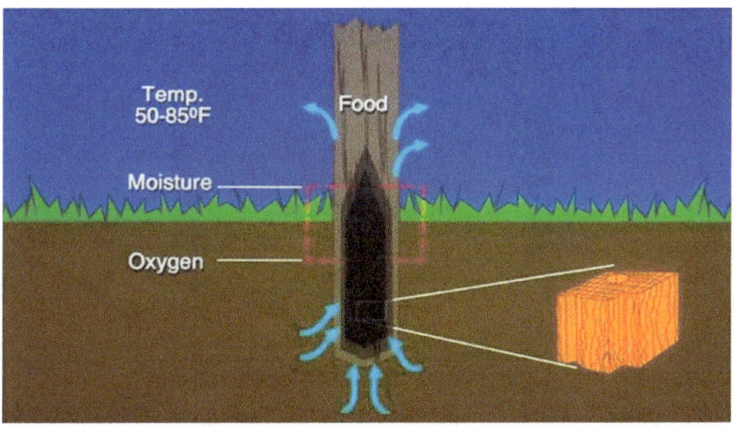

Figure 9-2. Key factors for decay to occur in wood pole structures. Source: Osmose Utilities Services, Inc., reprinted with permission.

species that enable decay to occur at greater depths. Therefore, the groundline zone of a pole where decay is most likely extends from 18 in. belowground to 6 in. aboveground. Below 18 in. the pole is usually in good condition, so maintaining the groundline zone extends pole life significantly.

9.3.1.1 Thin Sapwood Species. Wood in a standing tree is classified as either heartwood or sapwood. Sapwood is the living tissue in the outer portions of the tree that conducts water from the soil to the foliage above. Heartwood is a nonliving inner core of the tree. Different species have varying thicknesses of sapwood, depending on how the individual species conducts moisture. Heartwood and sapwood do not differ markedly in strength, but they do differ in their receptivity to preservative treatments. Sapwood is generally receptive to treatment, whereas heartwood is generally resistant.

Douglas fir and Western red cedar are thin sapwood species, as shown in Figure 9-3. The penetration of the original preservative treatment in these species is generally limited by sapwood thickness. Even though heartwood is naturally more resistant to decay than the sapwood, the heartwood may begin to decay when exposed to the proper conditions (as previously described).

Checks develop in thin sapwood species as poles season and dry while in service and tend to open beyond the treated sapwood, exposing the nontreated heartwood to fungal and insect attack as in Figure 9-4. The result is internal degradation, as depicted in Figure 9-5 and Figure 9-6. Internal decay in Douglas fir tends to occur in eccentric pockets, whereas Western red cedar more often develops concentric internal decay. This damage can continue to progress until only the preservative-treated sapwood shell remains.

Figure 9-3. Cross-section of Douglas fir showing a thin band of sapwood surrounding a heartwood core.
Source: Osmose Utilities Services, Inc., reprinted with permission.

Figure 9-4. Pole with very large check that extends to the center of the pole. Source: Osmose Utilities Services, Inc., reprinted with permission.

Figure 9-5 (top). Example of Douglas-fir poles with a dispersed decay pocket.

Figure 9-6 (bottom). Decay void in the heartwood. Note also the relatively shallow preservative-treated shell (dark color).
Source: Osmose Utilities Services, Inc., reprinted with permission.

9.3.1.2 Thick Sapwood Species. Southern pine has a thick sapwood that readily accepts preservatives, resulting in treatment of most of the critical pole cross-section. Figure 9-7 shows a cross-section view of a typical Southern pine.

As Southern pine poles age in service, the level of the active ingredients in the preservative may reduce to below-threshold values in the groundline zone. Eventually, fungi will begin to invade this wood and cause surface decay. This type of external decay is called "shell rot" and can rapidly cause a significant reduction in the bending capacity of a pole. See Figures 9-8 and 9-9 for examples of external decay.

9.3.2 Groundline Insect Damage

Fungal decay is by far the most common and critical issue for wood poles, but insects can also be a significant concern in some locations. The two most problematic insects for poles are carpenter ants and termites. Both excavate the interior of the poles, although they do so for different reasons. Carpenter ants remove sections of wood to create living space and do not use wood for food. These insects are more common in lines passing through forested areas. The nests in poles are often secondary or satellite nests. Thus, careful right-of-way maintenance to avoid the presence of

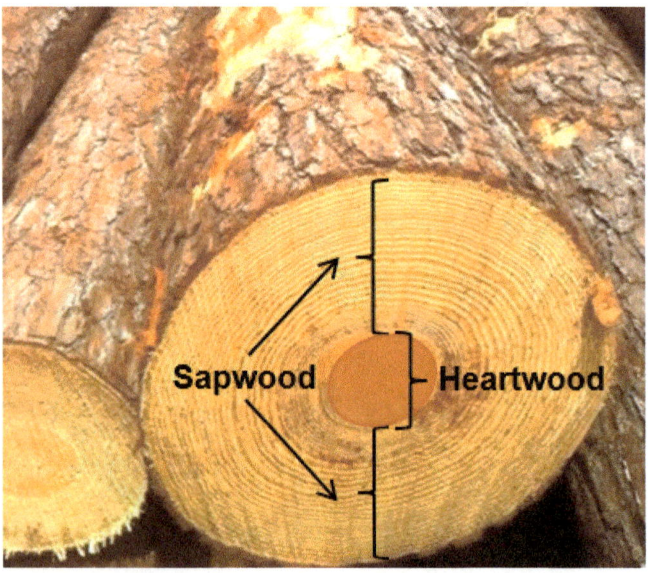

Figure 9-7. Southern pine tree showing the thick band of treatable sapwood on the cross-section, surrounding a small heartwood core.
Source: Osmose Utilities Services, Inc., reprinted with permission.

Figure 9-8 (top). Chipping to remove early-stage shell rot.

Figure 9-9 (bottom). Example of advanced shell rot.
Source: Osmose Utilities Services, Inc., reprinted with permission.

large woody debris that might harbor the main nest can help reduce the risk of damage.

Termites consume wood as their food source and are rarely seen outside the nest. Subterranean termites live in wood that is in or near the soil, whereas drywood termites live in wood that is near ground level as well as above the ground.

Termites and carpenter ants both create internal voids that can weaken a pole. Termites can also bring moisture and fungi into the pole, accelerating the degradation process. The impact on the pole's strength is evaluated by determining the extent of internal damage from these insects. Measurements are used to calculate the remaining section modulus, which determines the remaining bending strength of the pole.

9.3.3 Aboveground Degradation

An additional area of concern for wood pole decay is above the groundline at locations where field drilling has been used for connection hardware. Treating field-drilled bolt holes with a liquid preservative containing copper or copper and boron is recommended, to protect against decay in those locations.

Another aboveground decay concern is at the pole tops. All unprotected pole tops may eventually develop deep checks that allow moisture and decay fungi to enter into untreated wood. A pole cap can be added to prevent water from intruding into the pole (as mentioned in Chapter 8).

9.3.4 Woodpecker Damage

Woodpeckers do not use wood for food, but they cause damage aboveground when they excavate cavities for nests. This weakens the pole and exposes the pole interior to moisture and eventually fungal and insect attack. Options for preventing woodpecker damage include wrapping poles in wire mesh or using specific preservative treatments that purport to limit attack. Options for repairing woodpecker holes are discussed in Section 9.8 on Restoration.

9.3.5 Lightning/Electrical Damage

The frequency of lightning strikes varies greatly with geographical location. Lightning may cause burn damage near the top of the pole or cause the pole to split or burst if it contains sufficient moisture. In addition, wood poles can experience burn damage from electrical faults when the bonding and/or grounding of the electrical facilities are insufficient. A visual inspection from the groundline should identify lightning and electrical damage so that a utility crew can be mobilized to closely inspect and evaluate the full extent of the damage.

9.4 TRADITIONAL WOOD POLE INSPECTION AND ASSESSMENT PRACTICES

Since decay most often occurs belowground or internal to the pole, the inspection process needs to provide insight to the pole's condition at these locations. A variety of traditional tools and processes can be combined into an effective inspection program and used to improve the ability to identify and quantify decay conditions. Excavation around a pole provides the greatest opportunity for effective and reliable inspection and assessment. During this step, supplemental preservative paste can be applied as a booster shot to the original treatment for effective pole life extension.

The following sections describe the procedures and tools that make up the multiple steps of traditional wood pole inspection. Pole owners must configure a program specific to their assets based on pole species, decay hazard zone, years in service, previous maintenance efforts, goals for pole life extension, and other factors. In addition to groundline inspections and patrols, most transmission line owners also perform inspections of the upper portions of poles. Issues such as loose or missing hardware, cracked insulators, contaminated insulators, lightning damage, and conductor or splice damage can better be identified and evaluated using climbing or aerial inspections.

9.4.1 Aboveground Inspection

9.4.1.1 Visual Inspection. Visual inspection of wood pole structures is the first step of the assessment process whether conducting a walking patrol, groundline inspection, or climbing or aerial inspection. Effective visual inspection procedures are intended to identify more obvious conditions and maintenance-related issues such as broken crossarms or braces, missing nuts and bolts, and cracked insulators. The inspector also looks for damage to the pole itself, such as woodpecker holes and deteriorated tops. In addition, there may be subtle signs of insect damage such as bore holes, mounds of sawdust (frass), or mud tubes that warrant further assessment.

9.4.1.2 Climbing and Bucket Truck Inspections. Lineworkers climb the pole or use a bucket truck to perform a more detailed inspection of the aboveground issues. In particular, with X-braced H-frames, it is important to make sure the attachment hardware is securely in place and there is no damage to the braces. As they climb, line personnel may also use a hammer to sound the pole to detect any large internal voids. Figure 9-10 shows a lineworker performing an inspection from a bucket truck.

9.4.1.3 Aerial Inspections. Another option for aboveground inspections is using a helicopter or unmanned aerial system (UAS) to perform a flyby of the lines. Utility personnel either fly with a helicopter pilot and/or crew or a UAS captures images to be reviewed instantly or at a later time. When detailed inspections are required by the owner, the aircraft hovers around each structure and takes high-definition images of any structure, equipment, or conductor issues identified.

Figure 9-10. Bucket truck inspection.
Source: Intec Services, Inc., reprinted with permission.

9.4.2 Groundline Inspection

Groundline inspection uses visual, sounding, and boring techniques to evaluate the extent of internal decay (EPRI 2011). Some degree of excavation is also recommended to improve the accuracy of identifying decay conditions. Thick sapwood species, such as Southern pine, typically experience degradation externally belowground. Thin sapwood species such as Douglas fir and Western red cedar also experience degradation belowground, but it is more likely to occur internally. Partial or full circumferential excavation helps to reliably identify the presence and extent of belowground decay.

9.4.2.1 Visual and Sounding. Sounding with a hammer typically follows the visual inspection. The pole is hammered around ground level, below if excavated, and upward as high as the inspector can reach and around the entire circumference, as shown in Figure 9-11. The inspector listens for the dull thud that is indicative of internal decay pockets or insect damage in the pole. This practice is effective for detecting advanced internal decay, but it does not detect early stages of decay or insect damage.

9.4.2.2 Visual with Sound and Bore. The inspector sounds the pole to identify suspect areas and drills (bores) one or more holes of 3/8 in. to

Figure 9-11. Sounding inspection.
Source: Intec Services, Inc., reprinted with permission.

1/2 in. diameter to evaluate the presence and extent of the degradation. The holes are drilled at a steep slope (45 to 60 degrees) to intersect as much wood as possible, as shown in Figure 9-12. Drilling will detect internal decay pockets and some early decay but will not detect the earliest stage of incipient decay. In the case of thin sapwood species poles, larger inspection holes up to 7/8 in. in diameter can be drilled to enable inspection and also to serve as remedial treatment application holes.

9.4.2.3 Shell Thickness Indicator. Used in conjunction with drilling, a shell thickness indicator is a graduated metal rod with a hook at the end. The rod is inserted into the drilled holes and then pulled back along the hole. The hook catches voids or other internal defects, as pictured in Figure 9-13. The thickness of sound wood (or residual shell) can be measured along with the depth of the void. Those measurements are used to calculate the remaining pole-bending strength.

9.4.2.4 Excavating. A full excavation extends around the entire circumference to a depth of 18 in. in wetter climates and sometimes to a depth of up to 24 in. in drier climates. Excavation exposes the section of the pole where decay is most likely to occur.

In some cases, partial excavations are specified. This process varies according to the pole species and program goals. Methods of partial excavate

Figure 9-12. Drilling inspection.
Source: Intec Services, Inc., reprinted with permission.

Figure 9-13. Shell thickness indicator.
Source: J. Morrell, Oregon State University, reprinted with permission.

programs range from pulling back a small amount of soil to excavating one quadrant of the pole to a depth of 6 or 8 in., to excavating two quadrants of the pole to a depth of 8 in. in wetter climates, or digging one quadrant to a depth of 18 in. to 24 in. in dryer climates. If the excavation reveals surface decay or boring indicates internal decay, the pole may be further excavated around its circumference.

Excavation is effective for identifying both internal and external degradation below the ground that could otherwise be missed, and it affords an opportunity to externally treat the belowground portion of the pole to arrest existing degradation and resist future degradation.

9.4.2.5 Surface Scraper. A surface scraping tool, shown in Figure 9-14, is used to remove dirt belowground to allow the pole surface to be accurately assessed. The pointed tip of the triangular scraper is also useful for sensing early stages of shell rot.

9.4.2.6 Chipping Tool. This tool is used to remove decayed and weakened wood from the pole surface belowground, as shown in Figure 9-15. The residual circumference can then be accurately measured to estimate

Figure 9-14. Surface scraping tool.
Source: Osmose Utilities Services, Inc., reprinted with permission.

Figure 9-15. Chipping tool.
Source: Osmose Utilities Services, Inc., reprinted with permission.

the remaining strength. The decayed wood is removed since it is considered to have little or no structural integrity and can prevent remedial treatments from penetrating the remaining sound wood.

9.5 SERVICEABILITY DECISIONS

These traditional inspection methods have been widely used for more than 60 years. The objective is to utilize tools and implement procedures in a way that provides the highest probability of identifying and reliably measuring decay and other damage, while limiting the probability that damage or decay goes undetected and untreated. Measurements of damage are then used to calculate the remaining section modulus of the pole and assess the percentage of remaining bending strength. The strength of the wood is assumed to be the designated mean fiber strength, as discussed and defined in Chapter 3 of this manual. The pass/fail (serviceable/reject) determinations of the traditional inspection process have a long-established history of reliability.

Unlike most safety code requirements for steel or concrete structures, safety codes for wood poles in the United States and Canada allow for some level of deterioration before requiring structural rehabilitation or replacement. Usually a significant portion of the poles in a pole plant are found to be without decay.

A key element of an inspection program is the criteria used to determine if a weakened pole meets code strength requirements. Decayed poles that have adequate remaining strength are often called "decayed but serviceable"; those with less strength than required are termed "rejects."

The National Electrical Safety Code C2 (IEEE 2017, footnote in Table 261-1) makes a specific provision for an allowable amount of wood pole deterioration before a pole should be replaced. Similar statements are presented in the California Public Utilities Commission's "Rules for Overhead Line Construction," General Order 95 (GO 95) (California PUC 2018) and the Canadian Standards Association's CAN/CSA-022.3 No.1-010 rule, "Overhead Systems" (CSA 2001).

9.6 ADDITIONAL NONDESTRUCTIVE EVALUATION INSPECTION TOOLS AND PROCESSES

There have been many efforts to develop enhanced inspection instruments. The technologies explored include sonics, ground-penetrating radar, drilling resistance, mechanical penetration, vibration characteristics, and load-deflection measurements (CEATI International 2001). Each of these techniques seeks to measure some aspect of the effects of degradation. Care must be taken when considering the implementation of these methods to determine exactly what they measure and if the results are sufficiently accurate.

The goal for evaluating new technologies is to determine if they increase the value and effectiveness of inspections relative to traditional methods, provide more or better information, and/or lower inspection costs. For details about nondestructive testing (NDT) technologies, see Appendix D.

9.7 REMEDIAL TREATMENTS

Once the pole has been inspected and the remaining pole strength has been assessed, the next step is to consider future protection for the pole. Most utilities recognize their responsibility for maximizing the original pole investment through application of various remedial treatments to arrest or prevent internal and external decay.

External remedial treatment systems are normally applied to the below-ground portion of a pole after excavation. These treatments typically contain diffusible and nondiffusible preservatives such as water-soluble boron and oil-soluble copper. The water-soluble boron diffuses through the sapwood to control fungi established inside the poles, whereas the non-diffusible copper stays near the surface and limits reinvasion by fungi. These systems are usually applied on an 8- to 12-year cycle beginning when the pole is 10 to 20 years old. External systems typically penetrate more deeply into thick sapwood species poles. It is difficult for these systems to migrate into the heartwood of thin sapwood species poles.

There are a number of internal remedial treatment systems. Fumigants can be applied in liquid, granular, or solid forms that become gaseous and migrate up and down 2 to 3 ft from the point of application to sterilize sound wood. Fumigants are typically applied beginning at or below the groundline in a series of holes that extend upward in a spiral pattern.

Boron rods also provide decay protection for sound wood. The boron diffuses when water is present in the wood and moves through the wood away from the point of application. Although boron is an effective preservative treatment, it does not travel as far from the application point as fumigants, since it relies on moisture to carry it into the wood.

Boron rods and fumigants can be added to the inspection holes when larger inspection holes are used. All inspection and treatment holes are plugged. These treatments are normally applied on an 8- to 12-year cycle. It should be noted that boron rods and granular fumigants are moisture dependent and move at differing rates depending on available moisture.

Liquid void preservative treatments, as the name implies, are applied to voids in a pole. The goal is to coat the interior of the decay pocket or insect cavity with a preservative formulation that limits further damage. These liquid formulations contain one or more biocides and are applied under pressure through a series of bored holes 3/8 or 1/2 in. in diameter.

The American Wood Protection Association Standard M13 (AWPA 2015) can be referenced for additional information on remedial treatments.

9.8 RESTORATION

A majority of poles with remaining strength below code requirements are restorable. The steel truss method shown in Figure 9-16 is the most widely used restoration system and has been used for more than 50 years. It is typical for restored poles to remain safely in service for 30 or more years. The steel truss system provides significant savings over pole replacement, and many utilities are able to capitalize these costs.

Figure 9-16. Steel truss for restoration.
Source: Intec Services, Inc., reprinted with permission.

As with any type of groundline pole restoration system, a restored pole should be reinspected and retreated as part of the regular inspection and maintenance program. It is important not to overlook the need to keep decay from causing additional damage to the remaining wood.

Another option for restoring strength to a degraded or damaged section is to use a fiberglass wrap system, as shown in Figure 9-17. The process begins with an approximate 3 ft excavation, if near-ground deterioration, and application of a remedial paste treatment to help control future decay. The pole restoration product is then applied. A variety of fillers, fiberglass wraps, and steel splints are designed to restore strength aboveground when there is woodpecker or mechanical damage.

Data on the performance of these systems indicate that the combination of remedial treatment and pole restoration results in poles that are still serviceable 30 years after restoration. Restoration is a simple method for cost-effectively prolonging pole service life.

Figure 9-17. Fiberglass wrap installation.
Source: Intec Services, Inc., reprinted with permission.

REFERENCES

AWPA (American Wood Protection Association). 2015. M13, *Guidelines for a pole maintenance program*. Birmingham, AL: AWPA.

California PUC (Public Utilities Commission). 2018. *Rules for overhead line construction*, General Order No. 95. Accessed November 18, 2018. http://www.cpuc.ca.gov/gos/GO95/go_95_startup_page.html.

CEATI International (Centre for Energy Advancement through Technological Innovation). 2001. "Validating NDE tools for evaluating present condition, residual strength, and remaining life of in-service wood poles." *Tech. Rep. No. T004700-5005*. Montreal, QC: CEATI.

CEATI International. 2010. "Best practices for transmission line inspections and recommended inspection techniques." *Tech. Rep. No. T093700-3225*. Montreal, QC: CEATI.

CSA (Canadian Standards Association). 2001. CAN/CSA-022.3 No. 1-01, *Overhead systems*. Mississauga, ON: CSA Group.

EPRI (Electric Power Research Institute). 2004. *Wood pole management—issue assessment*. Palo Alto, CA: EPRI.

EPRI. 2005. *Evaluation of wood pole condition assessment tools*. Palo Alto, CA: EPRI.

EPRI. 2011. *Evaluating wood pole inspection alternatives to conventional visual, sound and bore practices*. Palo Alto, CA: EPRI.

IEEE (Institute of Electrical and Electronic Engineers). 2017. ANSI Standard C2, *National electrical safety code*. New York: IEEE.

Morrell, J. J. 1996. *Wood pole maintenance manual*. Special Publication 15. Corvallis, OR: Oregon State Univ., Forest Research Laboratory.

Morrell, J. J. 2012. *Wood pole maintenance manual*. Research Contribution 51. Corvallis, OR: Oregon State Univ., Forest Research Laboratory.

NEETRAC (National Electric Energy Testing Research and Applications Center). 2008. "Wood pole in-service assessment methods." *Baseline Project No. 05-250*. Forest Park, GA: NEETRAC.

Nelson, R. F. 2003. "Reliability-centered power line management: Inspection process, measurement techniques, and data management considerations." *Overhead line management and inspection workshop*. Fort Collins, CO: EDM International.

USDA (US Department of Agriculture). 2013. "Wood pole inspection and maintenance." *RUS Bull. 1730B-121*. Washington, DC: Government Printing Office.

APPENDIX A

RESILIENCY OF WOOD POLE OVERHEAD LINES

A.1 INTRODUCTION

Overhead electrical power lines, many of which are supported on wood poles, are an integral element of the US electrical grid. The electrical grid is a vital component of the nation's critical infrastructure. A more reliable, efficient, secure, and resilient electrical grid will improve service for all who rely on the grid for reliable electric power. This document defines resiliency as the ability of a system to withstand extreme weather events and to minimize the number, extent, and duration of outages.

A.2 RESILIENCE OF OVERHEAD POWER LINES

Because overhead power lines can never be 100% fail-safe, line owners should develop a plan that defines how best to strengthen and maintain the resiliency of their facilities. An effective resiliency plan will identify the greatest risks to the system and determine the cost and impact of the various efforts that can be implemented to minimize those risks. A cost-benefit analysis will then determine the level of effort that should be expended to efficiently improve and maintain the line resiliency.

Inspection and maintenance programs are the primary efforts employed by utility companies to sustain the resiliency established at the time of the initial wood pole installation. Any reduction of strength in the groundline zone over time results in a reduction in resiliency. Therefore, these programs start by inspecting wood poles on a regular cycle to evaluate and monitor their condition in the groundline zone where decay is most prevalent. Maintenance efforts include the application of supplemental preservative treatments that provide a booster shot to the original treatment to resist decay for an extended cycle. For decayed poles that are found to have a remaining strength that is below code requirements, restoration systems can be installed to strengthen the pole and avoid replacing the structure.

Increasing the resilience of overhead lines can also be planned for and performed during design. Weather event hardening can be achieved by using records of weather conditions from larger mean recurrence periods, such as 100-year or 200-year storm events, or by establishing other design load conditions that exceed those required by code. Material strength factors can also be adjusted such that the ultimate strength of the overhead line is increased and the line resiliency is inherently improved.

The resilience of in-service wood poles can be enhanced by installing extended steel trusses (see Figures A-1 and A-2) or other strength improvement systems. These systems are designed to function with a sound wood pole to increase the structure strength and improve the overall line resiliency.

Additional efforts commonly used to improve the resiliency of overhead lines include right-of-way inspection and maintenance and the practice of maintaining a supply of poles and standard line hardware for faster recovery/reconstruction of downed lines.

Some utility companies have implemented a program to replace their wood pole structures with steel, concrete, or other pole materials as a means of improving the resiliency of their overhead lines. However, changing the structure material alone is not an effective hardening strategy. The design criteria for loading and/or strength must also be changed in order to significantly improve line resiliency through structure replacement.

There are multiple aspects of overhead lines that can be addressed to improve resiliency. Utility companies should perform a comprehensive study of their facilities, systems, and environment to best prepare a plan that will effectively and efficiently strengthen and maintain the resiliency of their overhead lines.

A.3 IMPACT OF SEVERE WEATHER ON THE US ELECTRIC GRID

Severe weather is the leading cause of power outages in the United States. Outages caused by severe weather such as thunderstorms, hurricanes, ice storms, and blizzards account for more than 50% of outages observed since 2002 and more than 80% of outages affecting 50,000 or more customers (USDOE 2018). Figure A-3 indicates that the impact of weather-related events on observed outages has been growing in recent years.

Because of their statistical coefficient of variation (COV), round wood poles have a wide strength probability distribution. This is evident in the

Figure A-1 (left) and Figure A-2 (right). Upgrading steel truss during and after installation.
Source: Osmose Utilities Services, Inc., reprinted with permission.

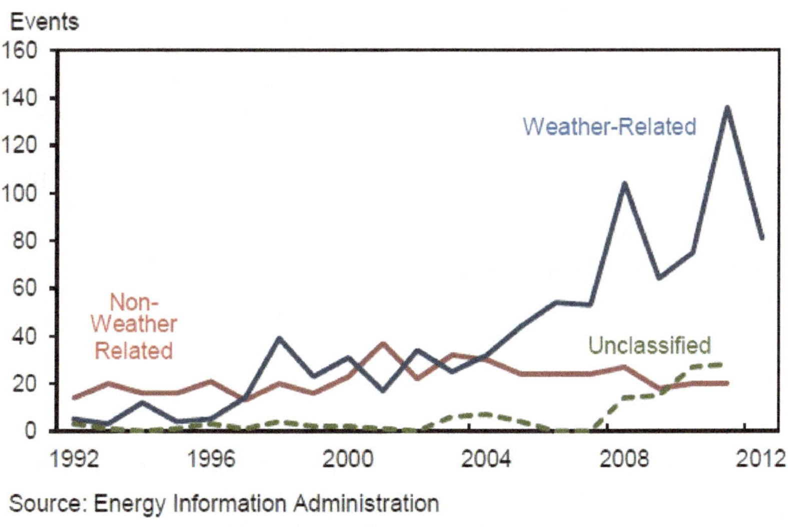

Figure A-3. Observed outages to the bulk electric system, 1992–2012.
Source: Economic Benefits of Increasing Electric Grid Resilience to Weather Outages, Executive Office of the President, August 2013.

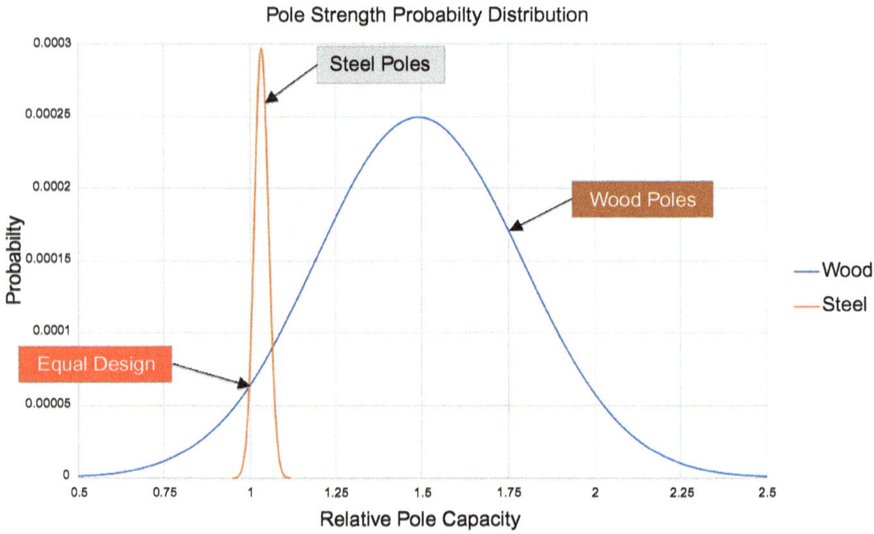

Figure A-4. Pole strength probability distribution.
Source: Otto Lynch, reprinted with permission.

chart in Figure A-4, with the strength distribution plotted for both typical steel and wood poles.

The point of equal design in Figure A-4 denotes the 5% Lower Exclusion Limit (LEL) of steel and wood poles. Using this 5% LEL as a basis for design for a given loading (e.g., wind, ice, etc.) allows all materials (e.g., wood, steel, concrete, fiberglass composites, etc.) to be compared, strength-wise, on an equal basis. As demonstrated graphically, round wood poles, owing to their statistical COV, have a wider strength probability distribution compared to steel poles. The significance of this statistical variation in the strength of a given material is that the greater variation in strength of the individual structures within a given line, the greater the resiliency of the line in its entirety to resist extreme loading events above the design loading.

A.4 SUMMARY

The use of wood pole structures, properly designed using consistent structural engineering principles, may provide a simple, cost-effective, and more resilient option than some of the other pole materials commonly used.

REFERENCE

USDOE (US Department of Energy). 2018. "Electric emergency incident and disturbance report." *Form OE-417*. Accessed November 16, 2018. https://www.oe.netl.doe.gov/oe417.aspx.

APPENDIX B
TRANSMISSION POLE DESIGN EXAMPLES

HAND CALCULATIONS

Knowns and Assumptions

Wood Pole Species: Douglas fir
Modulus of Rupture = 8,000 psi (see Table 1 of ANSI O5.1)
Coefficient of Variation = 0.2 (see Table 1, footnote 4 of ANSI O5.1)
Modulus of Elasticity = 2.38 ksi (see Table 1 of ANSI O5.1)
ANSI Class: H1
Total Length Pole (L): 75 ft
Embedment Depth (E): 9.5 ft (Calculated Embedment Depth, see Chapter 6)
Minimum Tip Circumference (C_t): 29 in. (see Table 8 of ANSI O5.1)
Minimum Circumference 6 ft from Butt (C_b): 55.5 in. (see Table 8 of ANSI O5.1) Density of Wood (ρ): 60 lb/ft^3

Calculations

Minimum Pole Tip Diameter:

$$D_t = \left(\frac{C_t}{\pi}\right) = \left(\frac{29}{\pi}\right) = 9.23 \text{ in.}$$

Minimum Diameter 6 ft from Butt:

$$D_b = \left(\frac{C_b}{\pi}\right) = \left(\frac{55.5}{\pi}\right) = 17.67 \text{ in.}$$

Pole Taper:

$$T = \left(\frac{(C_b - C_t)}{(L-6)}\right) = \left(\frac{(17.67 - 9.23)}{(75-6)}\right) = 0.122 \text{ in./ft}$$

Minimum Diameter at Groundline:

$$D_{GL} = D_t + (L-E)T = 9.23 + (75-9.5) * 0.122 = 17.24 \text{ in.}$$

Area at Groundline:

$$A_{GL} = \left(\frac{\pi D_{GL}^2}{4}\right) = \left(\frac{\pi * 17.24^2}{4}\right) = 233 \text{ in.}^2$$

Section Modulus at Groundline:

$$S_{GL} = \left(\frac{\pi D_{GL}^3}{32}\right) = \left(\frac{\pi * 17.24^3}{32}\right) = 503 \text{ in.}^2$$

Vertical Load at Groundline:

$$P_{GL} = \rho\left(\frac{(A_t) + (A_{GL})}{2(144)}\right)(L-E) = 4093 \text{ lbs}$$

Axial Stress at Groundline:

$$f_a = \left(\frac{P}{A}\right) = \left(\frac{4093}{233}\right) = 18 \text{ psi}$$

Moment at Groundline from Transverse Loads:

$$M_t = \Sigma(T * L) = (265 * 64.5) + (792 * 59.5) + (792 * 54.5) + (792 * 49.5)$$
$$= 146{,}585 \text{ ft-lbs}$$

Moment at Groundline from Vertical Loads:

$$M_v = \Sigma(V * e) = (137 * 0.5) + (546 * 5.5) - (546 * 5.5) + (546 * 5.5)$$
$$= 3{,}072 \text{ ft-lbs}$$

Moment at Groundline due to Wind on Pole:

$$M_{wp} = W_{STR} * \left(\frac{(2D_t + D_{GL})}{72}\right) * (L-E)^2$$
$$= 19.97 * \left(\frac{(2 * 9.23 + 17.24)}{72}\right) * 65.5^2 = 42{,}482 \text{ ft-lbs}$$

Total Linear Moment at Groundline:

$$M_{GLM} = M_t + M_v + M_{wp} = 146{,}585 + 3{,}072 + 42{,}482 = 192{,}139 \text{ ft-lbs}$$

P-Delta Amplification Factor

When using a first-order linear analysis to calculate a groundline moment (GLM) as shown, the deflected position of the structure after the loads are applied are not considered. This deflected position will result in additional moments to those calculated above that neglected the structure deflections. These deflections can be approximately calculated by several approximate hand methods. However, it is not a single process because once the first iteration is done, this will cause additional deflection, which will require another iteration. This process will need to be looped through many times until a final convergence is achieved. This is best done by nonlinear analysis software.

For the purposes of these example hand calculations, an approximate amplification factor using the Gere–Carter Method is calculated as

Vertical Load at Groundline:

$$V_L = P_{GL} + V_{TOTAL} = 4093 + 137 + 546 + 546 + 546 = 5868 \text{ lbs}$$

Effective Diameter of Pole:

$$D_{eff} = (D_t * D_{GL} = (9.23 * 17.24)^{0.5} = 12.6 \text{ in.}$$

Average Moment of Inertia:

$$I_{eff} = \left(\frac{\pi D_{eff}^4}{64}\right) = \left(\frac{\pi * 12.6^4}{64}\right) = 1243 \text{ in.}^4$$

Critical Buckling Load:

$$P'_{cr} = \frac{(\pi^2 * MOE * I_{eff})}{(4(L-E)^2 * 144)} = \frac{(\pi^2 * 2380000 * 1243)}{(4(75-9.5)^2 * 144)} = 11{,}815$$

Correction Factor:

$$P^* = \left(\frac{D_{GL}}{D_t}\right)^{2.7} = \left(\frac{17.24}{9.23}\right)^{2.7} = 5.40$$

Critical Buckling Load:

$$P'_{cr} = P'_{cr} * P^* = 11{,}815 * 5.40 = 63{,}796 \text{ lbs}$$

Est. P-Delta Moment Amplification Factor:

$$P_{deltaest} = \left(\frac{1}{\left(1-\left(\frac{V_L}{P_{cr}}\right)\right)}\right) = \left(\frac{1}{\left(1-\left(\frac{5868}{63{,}796}\right)\right)}\right) = 1.101$$

Total P-Delta Moment at Groundline:

$$M_{P\text{-}\Delta} = M_{GLM} * P_{deltaest} = 192{,}139 * 1.101 = 211{,}619 \text{ ft-lbs}$$

Bending Stress at Groundline:

$$f_b = \left(\frac{M_{P-\Delta}}{S_{GL}}\right) = \left(\frac{211{,}619 * 12}{503}\right) = 5050 \text{ psi}$$

Combined Stress:

$$f_c = f_a + f_b = 18 + 5050 = 5068 \text{ psi}$$

Allowable Stress with Height Factor:

$$F_{GL} = F_S\left(1 - \frac{.3H}{(L-E)}\right) = 8000\left(1 - \frac{.3 * 0}{(75 - 9.5)}\right) = 8000 \text{ psi*}$$

*(*Note: This calculation is not necessary for groundline calculations but is shown in the event hand calculations are used for another point on the pole.)*

Minimum Allowable Fiber Strength (Resistance) at Groundline with 5% LEL:

$$MOR_{5\%} = MOR_{50\%}(1 - 1.645 * COV) = 8000 * (1 - 1.645(.2)) = 5368 \text{ psi}$$

Pole Strength Utilization at Groundline:

$$\text{Usage \%} = \frac{f_c}{MOR_{5\%}} = \frac{5068}{5368} = 94.4\%$$

These calculations are only valid at the groundline of the pole. They must be repeated at other elevations up the pole to find the point of maximum stress. This would be rather difficult to do by hand, so the calculations were automated in a spreadsheet presented on the following page. It should be noted that the point of maximum stress is at 19.5 ft aboveground and is approaching 100% pole utilization.

Following the spreadsheet are the results of a nonlinear analysis. The results are slightly different (lower) owing to the approximate methods used to calculate the deflection effects as shown. However, the point of maximum stress is very close to the same point as the hand calculation method.

Additional examples are presented on the following pages to demonstrate the structural analysis of four other structure configurations. Because of the complexity of the calculations for these structure types, only the nonlinear analyses are provided. Should hand calculations wish to be done for structures of those types, there are many historical documents that will provide approximate methods for doing so. One resource is *Transmission Line Design Manual* (Farr 1980).

REFERENCES

ANSI/AWPA (American Wood Protection Association). 2017. ANSI-O5.1, *Wood poles—specifications and dimensions*. Birmingham, AL: AWPA.

Farr, Holland H. 1980. *Transmission line design manual*. Washington, DC: Western Area Power Administration, US Department of the Interior.

USDA (US Department of Agriculture). 1998a. "Electric transmission specifications and drawings, 115 kV through 230 kV." *RUS Bull. 1728F-811*. Washington, DC: Government Printing Office. https://www.rd.usda.gov/files/UEP_Bulletin_1728F-811.pdf.

USDA. 1998b. "Electric transmission specifications and drawings, 34.5 kV through 69 kV." *RUS Bull. 1728F-810*. Washington, DC: Government Printing Office. https://www.rd.usda.gov/files/UEP_Bulletin_1728F-810.pdf.

USDA. 1998c. "Specifications and drawings for 24.9/14.4 kV line construction." *RUS Bull. 1728F-803*. Washington, DC: Government Printing Office. https://www.rd.usda.gov/files/UEP_Bulletin_1728F-803.pdf.

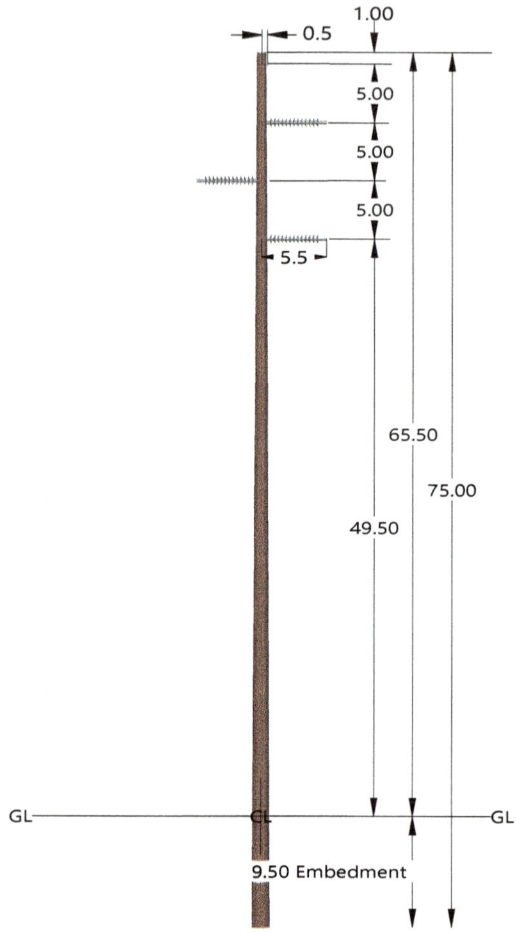

DESIGN PARAMETERS:	TRANSMISSION POLE DESIGN EXAMPLE 1
75\` H1 DOUGLAS FIR POLE WEIGHT OF WOOD 60 PCF MOR = 8,000 PSI Strength Factor = 0.671 MOE = 2,380,000 PSI	ASCE WOOD POLE STRUCTURE MANUAL OF PRACTICE

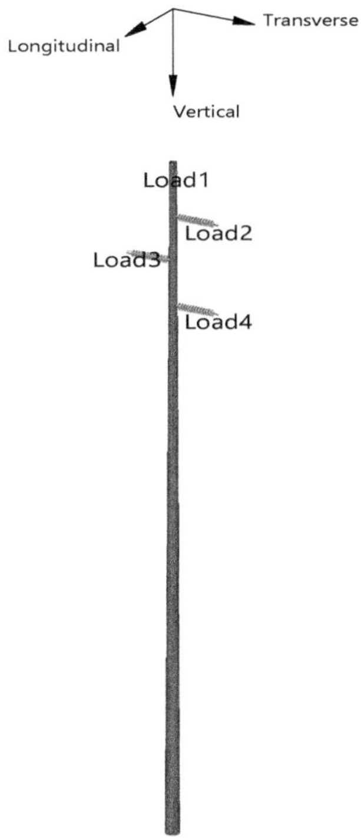

Loading Tree

Load Case	Joint Label	Vertical	Transverse	Longitudinal	Transverse Wind (psf)
		Load (lbs)			
ASCE	Load1	137	265	0	19.97
	Load2	546	792	0	19.97
	Load3	546	792	0	19.97
	Load4	546	792	0	19.97

Report Generated: 10:50:40 AM 8/11/2015

DESIGN PARAMETERS:	TRANSMISSION POLE DESIGN EXAMPLE 1
75` H1 DOUGLAS FIR POLE WEIGHT OF WOOD 60 PCF MOR = 8,000 PSI Strength Factor = 0.671 MOE = 2,380,000 PSI	ASCE WOOD POLE STRUCTURE MANUAL OF PRACTICE

Transmission Pole Design Example 1 - Typical Manual 74 Type loads (Load Tree)

Pole Selected for Analysis:

Species:	DF	(Douglas Fir = DF; Southern Yellow Pine =SYP, Western Red Cedar=WRC		MO
ANSI Class:	H1		(INPUT VALUE- ANSI 01.5 Class H6 through C10	C
Total Length Pole:	75	ft.	(INPUT VALUE)	MO
Embedment Depth:	9.5	ft.	(Calculated Embedment Depth - See Chapter 6)	MC
Minimum Tip Circumference:	29	in.	(Lookup From Table 8 of ANSI 05.1)	
Circumference 6ft from butt:	55.5	in.	(Lookup From Table 8 of ANSI 05.1)	Nominal Strength Adjustme
Density of Wood (ρ):	60	pcf	(INPUT VALUE)	

Minimum Tip Dia:	9.23	in.	(Calculated)
Minimum Dia 6ft from butt:	17.67	in.	(Calculated)
Pole Taper:	0.12225	in/ft	(Calculated)
Minimum Dia at Groundline:	17.24	in.	(Calculated)

Applied Loads:

	From Tip Dist	
Load 1	1	ft.
Load 2	6	ft.
Load 3	11	ft.
Load 4	16	ft.
Load 5	0	ft.
Wind	19.97	psf

Distance from Tip	Dia.	Area	S_x	Vertical Load	Axial Stress	Applied Moment From Transverse Load	Applied Moment Fr Eccentric Vertical L
(L)	(D)	$A=(\frac{\pi D^2}{4})$	$S_x=(\frac{\pi D^3}{32})$	$P=\rho * \left(\frac{\left(\frac{\pi D^2}{4}\right)}{144}\right) * L$	$f_a=(\frac{P}{A})$	$M_t=\Sigma(T * L)$	$M_e=\Sigma(V * e)$
(ft.)	(in)	(in^2)	(in^3)	lbs	(psi)	(ft-lbs)	(ft-lbs)
0	9.23	67	77	0	0	0	0
5	9.84	76	94	149	2	1060	69
10	10.45	86	112	318	4	5553	3072
15	11.06	96	133	510	5	14006	69
20	11.68	107	156	725	7	26419	3072
25	12.29	119	182	966	8	39624	3072
30	12.90	131	211	1235	9	52829	3072
35	13.51	143	242	1533	11	66034	3072
40	14.12	157	276	1863	12	79239	3072
45	14.73	170	314	2225	13	92444	3072
50	15.34	185	355	2623	14	105649	3072
55	15.95	200	399	3058	15	118854	3072
60	16.57	216	446	3531	16	132059	3072
65	17.18	232	498	4044	17	145264	3072
65.5	17.24	233	503	4098	18	146585	3072
65.5	17.24	233	503	4098	18	146585	3072
65.5	17.24	233	503	4098	18	146585	3072
65.5	17.24	233	503	4098	18	146585	3072
65.5	17.24	233	503	4098	18	146585	3072
65.5	17.24	233	503	4098	18	146585	3072
65.5	17.24	233	503	4098	18	146585	3072
65.5	17.24	233	503	4098	18	146585	3072
65.5	17.24	233	503	4098	18	146585	3072
65.5	17.24	233	503	4098	18	146585	3072
65.5	17.24	233	503	4098	18	146585	3072
65.5	17.24	233	503	4098	18	146585	3072

8,000	psi	(From Table 1 of ANSI 05.1)
0.20		(From Table 1, Footnote 4 of ANSI 05.1)
2,380,000	psi	(From Table 1 of ANSI 05.1)
5368	psi	(5% LEL - = MORm - 1.645(MORm*COV))
0.671	$=MOE_5/MORm$	

Eccentric Dist (e)		T Load (lbs)	V Load (lbs)
0.5	ft.	265	137
5.5	ft.	792	546
-5.5	ft.	792	546
5.5	ft.	792	546
0	ft.	0	0
			1775

Moment due to Wind on Pole	Total Moment w/ P-Delta Factor	Bending Stress	Combined Stress from Loads	Height Factor	MOR w/ Height Adjustment	5% LEL Strength (Resistance)	Strength Utilization	
M_w= Wpsf $* (\frac{2Dt+Db)}{72}) * L^2$	$M_{total}=P_{deltaest.} * (M_t+M_e+M_w)$	$f_b=(\frac{M_{tot}}{S_x})$	$R=(fa+fb)$	$HF=MOR*(1-\frac{0.3H}{L})$ (Max reduction 15%)	$MOR_{adj}=MOR_m*HF$	$R_5=MOR_{adj}*LELadj$	$(\frac{R}{R_{allow}})$	
(ft-lbs)	(ft-lbs)	(psi)	(psi)		Stress	(psi)	%	
0	0	0	0	0.85	6,800	4563	0.00%	
196	1459	187	189	0.85	6,800	4563	4.14%	
802	10382	1111	1115	0.85	6,800	4563	24.43%	
1843	17531	1582	1587	0.85	6,800	4563	34.78%	
3344	36163	2777	2784	0.85	6,800	4563	61.01%	
5330	52895	3485	3493	0.85	6,800	4563	76.56%	
7828	70191	3998	4007	0.85	6,800	4563	87.83%	
10863	88077	4366	4377	0.86	6,882	4618	94.78%	
14460	106582	4627	4639	0.88	7,066	4741	97.84%	
18644	125734	4807	4820	0.91	7,249	4864	99.09%	
23441	145562	4926	4940	0.93	7,432	4987	99.05%	
28876	166092	4999	5014	0.95	7,615	5110	98.13%	
34975	187354	5037	5054	0.97	7,798	5233	96.58%	
41764	209374	5049	5067	1.00	7,982	5356	94.61%	
42482	211619	5050	5067	1.00	8,000	5368	94.39%	Groundline
42482	211619	5050	5067	1.00	8,000	5368	94.39%	
42482	211619	5050	5067	1.00	8,000	5368	94.39%	
42482	211619	5050	5067	1.00	8,000	5368	94.39%	
42482	211619	5050	5067	1.00	8,000	5368	94.39%	
42482	211619	5050	5067	1.00	8,000	5368	94.39%	
42482	211619	5050	5067	1.00	8,000	5368	94.39%	
42482	211619	5050	5067	1.00	8,000	5368	94.39%	
42482	211619	5050	5067	1.00	8,000	5368	94.39%	
42482	211619	5050	5067	1.00	8,000	5368	94.39%	
42482	211619	5050	5067	1.00	8,000	5368	94.39%	
42482	211619	5050	5067	1.00	8,000	5368	94.39%	

Detailed Wood Pole Usages for Load Case "ASCE"

Element Label	Joint Label	Joint Position	Rel. Dist. (ft)	Trans. Defl. (in)	Long. Defl. (in)	Vert. Defl. (in)	Trans. Mom. (ft-k)	Long. Mom. (ft-k)	Tors. Mom. (ft-k)	Axial Force (kips)	Tran. Shear (kips)	Long. Shear (kips)	P/A (ksi)	M/S. (ksi)	Res. (ksi)	Max. Usage %
Pole	Pole:t	Origin	0.00	74.86	0.07	-4.61	-0.00	-0.00	0.0	-0.01	0.01	-0.00	0.00	0.00	0.00	0.0
Pole	Pole:Load1	End	1.00	72.97	0.06	-4.46	0.01	-0.00	0.0	-0.01	0.01	-0.00	0.00	0.00	0.00	0.0
Pole	Pole:Load1	Origin	1.00	72.97	0.06	-4.46	0.08	-0.00	-0.0	-0.20	0.35	-0.00	0.00	0.01	0.01	0.3
Pole	Pole:Load2	End	6.00	63.54	0.06	-3.71	1.85	-0.00	-0.0	-0.20	0.35	-0.00	0.00	0.23	0.23	5.1
Pole	Pole:Load2	Origin	6.00	63.54	0.06	-3.71	4.10	-0.00	-0.0	-0.78	1.33	-0.00	0.01	0.51	0.52	11.3
Pole	Pole:Load3	End	11.00	54.24	0.05	-2.99	10.73	-0.01	-0.0	-0.78	1.33	-0.00	0.01	1.11	1.12	24.5
Pole	Pole:Load3	Origin	11.00	54.24	0.05	-2.99	8.39	-0.01	-0.0	-1.38	2.30	-0.00	0.02	0.87	0.88	19.4
Pole	Pole:Load4	End	16.00	45.23	0.04	-2.31	19.90	-0.01	-0.0	-1.38	2.30	-0.00	0.01	1.74	1.75	38.4
Pole	Pole:Load4	Origin	16.00	45.23	0.04	-2.31	22.20	-0.01	-0.0	-2.03	3.27	-0.00	0.02	1.94	1.96	42.9
Pole	#Pole:0	End	21.00	36.72	0.03	-1.70	38.55	-0.02	-0.0	-2.03	3.27	-0.00	0.02	2.87	2.89	63.3
Pole	#Pole:0	Origin	21.00	36.72	0.03	-1.70	38.55	-0.02	-0.0	-2.30	3.37	-0.00	0.02	2.87	2.89	63.3
Pole	#Pole:1	End	26.00	28.93	0.03	-1.19	55.41	-0.03	-0.0	-2.30	3.37	-0.00	0.02	3.54	3.56	78.1
Pole	#Pole:1	Origin	26.00	28.93	0.03	-1.19	55.41	-0.03	-0.0	-2.60	3.47	-0.00	0.02	3.54	3.57	78.2
Pole	#Pole:2	End	31.00	22.00	0.02	-0.79	72.76	-0.05	-0.0	-2.60	3.47	-0.00	0.02	4.03	4.05	88.7
Pole	#Pole:2	Origin	31.00	22.00	0.02	-0.79	72.76	-0.05	-0.0	-2.93	3.56	-0.00	0.02	4.03	4.05	88.8
Pole	#Pole:3	End	36.00	15.99	0.02	-0.49	90.58	-0.07	-0.0	-2.93	3.56	-0.00	0.02	4.37	4.39	94.6
Pole	#Pole:3	Origin	36.00	15.99	0.02	-0.49	90.58	-0.07	-0.0	-3.29	3.65	-0.00	0.02	4.37	4.39	94.6
Pole	#Pole:4	End	41.00	10.94	0.01	-0.27	108.85	-0.09	-0.0	-3.29	3.65	-0.00	0.02	4.60	4.63	97.1
Pole	#Pole:4	Origin	41.00	10.94	0.01	-0.27	108.85	-0.09	-0.0	-3.68	3.74	-0.00	0.02	4.60	4.63	97.1
Pole	#Pole:5	End	46.00	6.87	0.01	-0.14	127.55	-0.11	-0.0	-3.68	3.74	-0.00	0.02	4.76	4.78	97.7
Pole	#Pole:5	Origin	46.00	6.87	0.01	-0.14	127.55	-0.11	-0.0	-4.10	3.82	-0.01	0.02	4.76	4.78	97.8
Pole	#Pole:6	End	51.00	3.76	0.00	-0.05	146.67	-0.14	-0.0	-4.10	3.82	-0.01	0.02	4.85	4.87	97.1
Pole	#Pole:6	Origin	51.00	3.76	0.00	-0.05	146.67	-0.14	-0.0	-4.56	3.90	-0.01	0.02	4.85	4.87	97.2
Pole	#Pole:7	End	56.00	1.60	0.00	-0.02	166.17	-0.17	-0.0	-4.56	3.90	-0.01	0.02	4.89	4.91	95.6
Pole	#Pole:7	Origin	56.00	1.60	0.00	-0.02	166.17	-0.17	-0.0	-5.03	3.97	-0.01	0.02	4.89	4.91	95.7
Pole	#Pole:8	End	60.75	0.40	0.00	-0.00	185.02	-0.20	-0.0	-5.03	3.97	-0.01	0.02	4.89	4.92	93.6
Pole	#Pole:8	Origin	60.75	0.40	0.00	-0.00	185.02	-0.20	-0.0	-5.51	4.03	-0.01	0.03	4.89	4.92	93.7
Pole	Pole:g	End	65.50	0.00	0.00	0.00	204.18	-0.23	-0.0	-5.51	4.03	-0.01	0.02	4.87	4.90	91.2

DESIGN PARAMETERS:	TRANSMISSION POLE DESIGN EXAMPLE 1
75\` H1 DOUGLAS FIR POLE WEIGHT OF WOOD 60 PCF MOR = 8,000 PSI Strength Factor = 0.671 MOE = 2,380,000 PSI	ASCE WOOD POLE STRUCTURE MANUAL OF PRACTICE

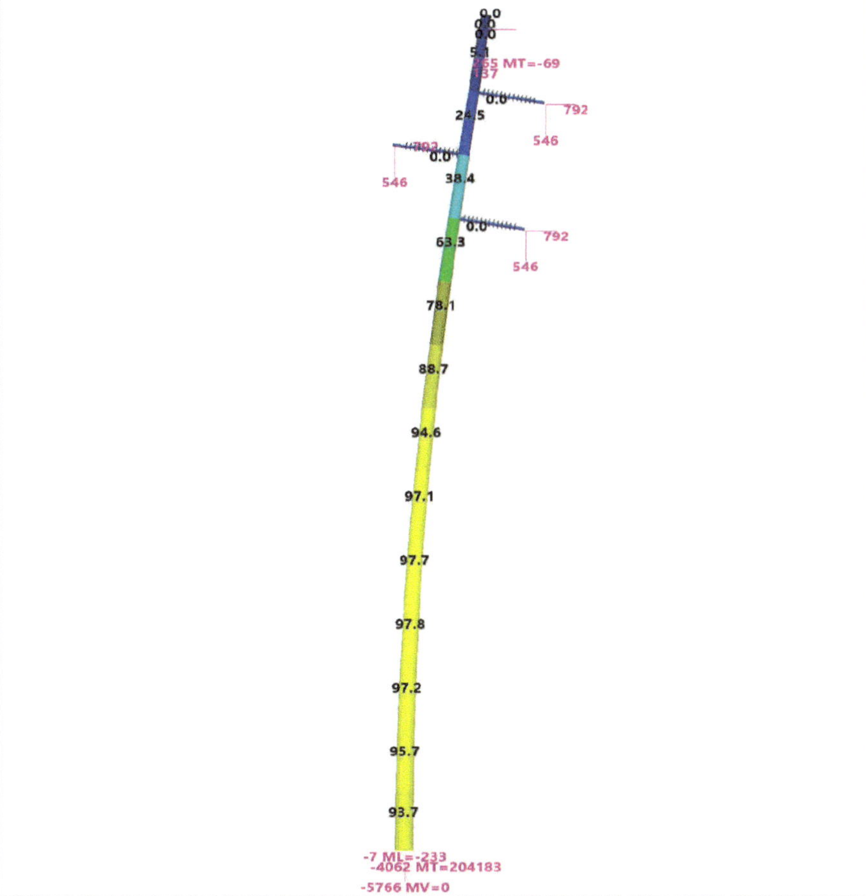

DESIGN PARAMETERS:	TRANSMISSION POLE DESIGN EXAMPLE 1
75` H1 DOUGLAS FIR POLE WEIGHT OF WOOD 60 PCF MOR = 8,000 PSI Strength Factor = 0.671 MOE = 2,380,000 PSI	ASCE WOOD POLE STRUCTURE MANUAL OF PRACTICE

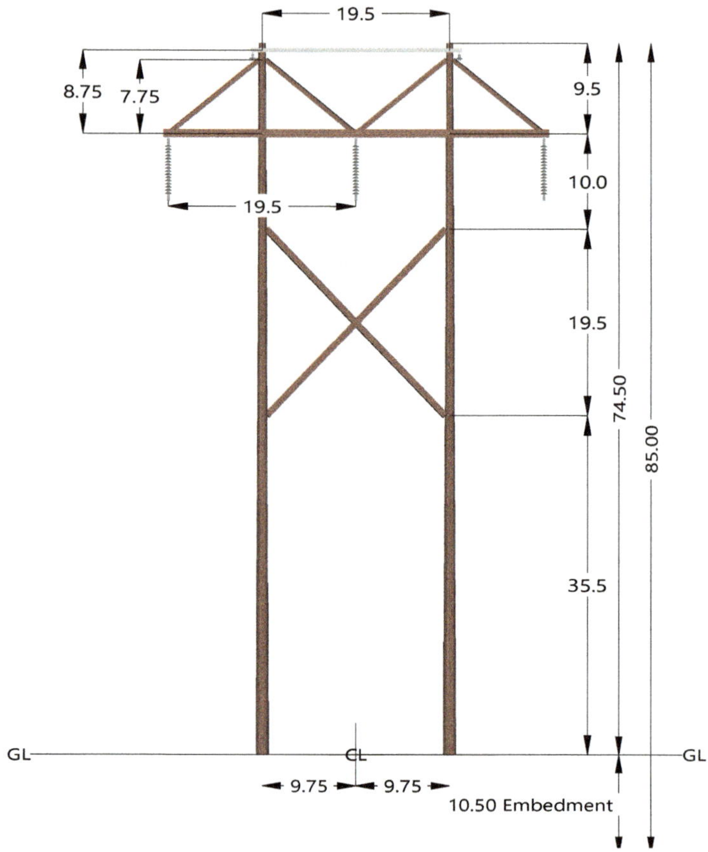

DESIGN PARAMETERS:	TRANSMISSION POLE DESIGN EXAMPLE 2
85` C2 SOUTHERN PINE POLES WEIGHT OF WOOD 60 PCF MOR = 8,000 PSI Strength Factor = 0.671 MOE = 2,130,000 PSI	ASCE WOOD POLE STRUCTURE MANUAL OF PRACTICE

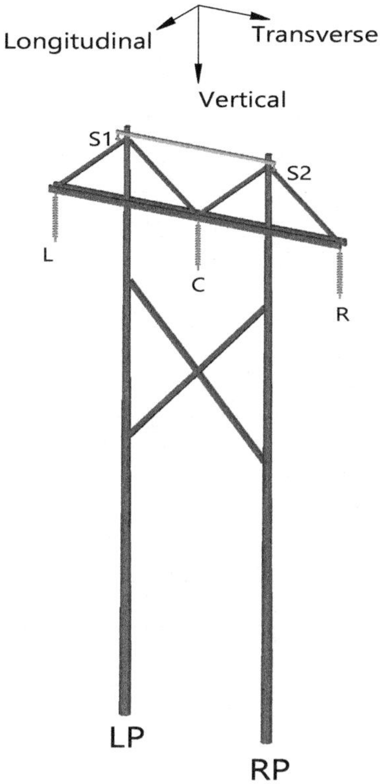

Loading Tree

Load Case	Joint Label	Vertical	Transverse	Longitudinal	Transverse Wind (psf)
		Load (lbs)			
ASCE Design Example 2	S1	1044	983	0	10
	S2	1044	983	0	10
	L	3143	1581	0	10
	C	3143	1581	0	10
	R	3143	1581	0	10

DESIGN PARAMETERS:

85` C2
SOUTHERN PINE POLES
WEIGHT OF WOOD 60 PCF
MOR = 8,000 PSI
Strength Factor = 0.671
MOE = 2,130,000 PSI

TRANSMISSION POLE DESIGN EXAMPLE 2

ASCE WOOD POLE STRUCTURE MANUAL OF PRACTICE

Detailed Wood Pole Usages for Load Case "ASCE Design Example 2,I NA+"

Element Label	Joint Label	Joint Position	Rel. Dist. (ft)	Trans. Defl. (in)	Long. Defl. (in)	Vert. Defl. (in)	Trans. Mom. (ft-k)	Long. Mom. (ft-k)	Tors. Mom. (ft-k)	Axial Force (kips)	Tran. Shear (kips)	Long. Shear (kips)	P/A (ksi)	M/S. (ksi)	Res. (ksi)	Max. Usage %
LP	LP:t	Origin	0.00	11.26	5.31	-0.17	0.00	-0.00	-0.0	-0.01	0.00	-0.00	0.00	0.00	0.00	0.0
LP	LP:XA	End	0.75	11.28	5.20	-0.17	0.00	-0.00	-0.0	-0.01	0.00	-0.00	0.00	0.00	0.00	0.0
LP	LP:XA	Origin	0.75	11.28	5.20	-0.17	-0.00	-0.54	1.2	-1.25	2.86	-0.04	0.03	0.16	0.19	4.2
LP	LP:KB	End	1.75	11.31	5.06	-0.17	2.86	-0.57	1.2	-1.25	2.86	-0.04	0.03	0.81	0.84	19.0
LP	LP:KB	Origin	1.75	11.31	5.06	-0.17	1.53	-0.57	1.2	-4.67	-1.83	-0.03	0.10	0.45	0.55	12.5
LP	#LP:0	End	5.62	11.41	4.51	-0.16	-5.56	-0.68	1.2	-4.67	-1.83	-0.03	0.08	1.09	1.17	26.6
LP	#LP:0	Origin	5.62	11.41	4.51	-0.16	-5.57	-0.68	1.2	-4.82	-1.78	-0.03	0.08	1.09	1.18	26.6
LP	LP:ARM	End	9.50	11.27	4.00	-0.16	-12.46	-0.80	1.2	-4.82	-1.78	-0.03	0.09	2.12	2.21	49.9
LP	LP:ARM	Origin	9.50	11.27	4.00	-0.16	-12.45	-0.70	0.7	-5.54	3.46	-0.08	0.10	2.12	2.22	50.2
LP	#LP:1	End	14.50	10.63	3.38	-0.15	4.84	-1.09	0.7	-5.54	3.46	-0.08	0.08	0.71	0.78	17.7
LP	#LP:1	Origin	14.50	10.63	3.38	-0.15	4.84	-1.10	0.7	-5.77	3.48	-0.07	0.08	0.71	0.79	17.8
LP	LP:XB1	End	19.50	10.20	2.81	-0.14	22.23	-1.45	0.7	-5.77	3.48	-0.07	0.08	3.20	3.28	74.2
LP	LP:XB1	Origin	19.50	10.20	2.81	-0.14	19.28	-1.46	0.7	1.02	-3.34	-0.06	-0.01	2.77	2.79	63.1
LP	#LP:2	End	24.50	10.39	2.29	-0.14	2.57	-1.76	0.7	1.02	-3.34	-0.06	-0.01	0.32	0.33	7.6
LP	#LP:2	Origin	24.50	10.39	2.29	-0.14	2.57	-1.76	0.7	0.75	-3.30	-0.07	-0.01	0.32	0.33	7.5
LP	#LP:3	End	29.50	10.70	1.82	-0.14	-13.92	-2.11	0.7	0.75	-3.30	-0.07	-0.01	1.26	1.26	28.6
LP	#LP:3	Origin	29.50	10.70	1.82	-0.14	-13.92	-2.10	0.7	0.44	-3.26	-0.07	-0.00	1.26	1.26	28.5
LP	#LP:4	End	34.25	10.66	1.44	-0.13	-29.39	-2.45	0.7	0.44	-3.26	-0.07	-0.00	2.30	2.30	52.1
LP	#LP:4	Origin	34.25	10.66	1.44	-0.13	-29.39	-2.44	0.7	0.09	-3.21	-0.08	-0.00	2.30	2.30	52.1
LP	LP:XB2	End	39.00	10.02	1.10	-0.13	-44.64	-2.81	0.7	0.09	-3.21	-0.08	-0.00	3.53	3.53	79.3
LP	LP:XB2	Origin	39.00	10.02	1.10	-0.13	-48.51	-2.80	0.7	7.40	4.13	-0.08	-0.07	3.84	3.91	87.7
LP	#LP:5	End	44.00	8.59	0.80	-0.11	-27.86	-3.19	0.7	7.40	4.13	-0.08	-0.06	1.69	1.75	38.4
LP	#LP:5	Origin	44.00	8.59	0.80	-0.11	-27.86	-3.19	0.7	7.04	4.14	-0.08	-0.06	1.69	1.75	38.3
LP	#LP:6	End	49.00	6.71	0.55	-0.09	-7.16	-3.61	0.7	7.04	4.14	-0.08	-0.05	0.43	0.48	10.3
LP	#LP:6	Origin	49.00	6.71	0.55	-0.09	-7.16	-3.61	0.7	6.62	4.20	-0.09	-0.05	0.43	0.48	10.2
LP	#LP:7	End	54.00	4.72	0.35	-0.05	13.81	-4.06	0.7	6.62	4.20	-0.09	-0.04	0.68	0.73	15.2
LP	#LP:7	Origin	54.00	4.72	0.35	-0.05	13.81	-4.06	0.7	6.14	4.28	-0.10	-0.04	0.68	0.72	15.1
LP	#LP:8	End	59.00	2.88	0.20	-0.03	35.23	-4.54	0.7	6.14	4.28	-0.10	-0.04	1.50	1.54	31.5
LP	#LP:8	Origin	59.00	2.88	0.20	-0.03	35.23	-4.55	0.7	5.62	4.39	-0.10	-0.04	1.50	1.53	31.5
LP	#LP:9	End	64.00	1.40	0.09	-0.01	57.19	-5.05	0.7	5.62	4.39	-0.10	-0.03	2.17	2.20	44.2
LP	#LP:9	Origin	64.00	1.40	0.09	-0.01	57.19	-5.06	0.7	5.04	4.51	-0.11	-0.03	2.17	2.20	44.1
LP	#LP:10	End	69.00	0.40	0.02	-0.00	79.73	-5.59	0.7	5.04	4.51	-0.11	-0.03	2.71	2.74	53.9
LP	#LP:10	Origin	69.00	0.40	0.02	-0.00	79.73	-5.59	0.7	4.57	4.60	-0.11	-0.02	2.71	2.74	53.9
LP	#LP:11	End	71.75	0.10	0.01	0.00	92.37	-5.89	0.7	4.57	4.60	-0.11	-0.02	2.97	2.99	58.2
LP	#LP:11	Origin	71.75	0.10	0.01	0.00	92.37	-5.89	0.7	4.21	4.66	-0.11	-0.02	2.97	2.99	58.2
LP	LP:g	End	74.50	0.00	0.00	0.00	105.20	-6.19	0.7	4.21	4.66	-0.11	-0.02	3.20	3.22	61.9
RP	RP:t	Origin	0.00	11.28	4.29	-0.24	0.00	0.00	-0.0	-0.01	0.00	-0.00	0.00	0.00	0.00	0.0
RP	RP:XA	End	0.75	11.32	4.18	-0.24	0.00	-0.00	-0.0	-0.01	0.00	-0.00	0.00	0.00	0.00	0.0
RP	RP:XA	Origin	0.75	11.32	4.18	-0.24	0.00	-0.53	-0.4	-1.14	-0.88	-0.01	0.03	0.15	0.18	4.1
RP	RP:KB	End	1.75	11.37	4.04	-0.24	-0.88	-0.54	-0.4	-1.14	-0.88	-0.01	0.03	0.27	0.30	6.7
RP	RP:KB	Origin	1.75	11.37	4.04	-0.24	-1.06	-0.54	-0.4	-8.47	-1.53	-0.02	0.19	0.32	0.51	11.5
RP	#RP:12	End	5.62	11.47	3.51	-0.23	-6.98	-0.62	-0.4	-8.47	-1.53	-0.02	0.15	1.37	1.51	34.2
RP	#RP:12	Origin	5.62	11.47	3.51	-0.23	-6.98	-0.62	-0.4	-8.62	-1.44	-0.02	0.15	1.37	1.52	34.3
RP	RP:ARM	End	9.50	11.26	3.01	-0.23	-12.56	-0.69	-0.4	-8.62	-1.44	-0.02	0.16	2.13	2.29	51.8
RP	RP:ARM	Origin	9.50	11.26	3.01	-0.23	-12.56	-0.77	0.6	-9.55	3.79	-0.11	0.17	2.13	2.31	52.2
RP	#RP:13	End	14.50	10.52	2.43	-0.22	6.41	-1.29	0.6	-9.55	3.79	-0.11	0.13	0.93	1.06	24.1
RP	#RP:13	Origin	14.50	10.52	2.43	-0.22	6.41	-1.30	0.6	-9.79	3.79	-0.10	0.14	0.93	1.07	24.2
RP	RP:XB1	End	19.50	10.06	1.90	-0.21	25.35	-1.76	0.6	-9.79	3.79	-0.10	0.14	3.64	3.78	85.6
RP	RP:XB1	Origin	19.50	10.06	1.90	-0.21	21.98	-1.77	0.6	-18.09	-3.69	-0.10	0.26	3.16	3.42	77.4
RP	#RP:14	End	24.50	10.34	1.45	-0.20	3.54	-2.27	0.6	-18.09	-3.69	-0.10	0.21	0.43	0.64	14.5
RP	#RP:14	Origin	24.50	10.34	1.45	-0.20	3.54	-2.27	0.6	-18.35	-3.69	-0.08	0.21	0.43	0.64	14.6
RP	#RP:15	End	29.50	10.75	1.07	-0.19	-14.91	-2.70	0.6	-18.35	-3.69	-0.08	0.19	1.35	1.54	34.9
RP	#RP:15	Origin	29.50	10.75	1.07	-0.19	-14.91	-2.69	0.6	-18.67	-3.53	-0.07	0.19	1.35	1.54	34.9
RP	#RP:16	End	34.25	10.79	0.77	-0.19	-31.68	-3.01	0.6	-18.67	-3.53	-0.07	0.18	2.48	2.66	60.1
RP	#RP:16	Origin	34.25	10.79	0.77	-0.19	-31.68	-3.00	0.6	-19.02	-3.27	-0.05	0.18	2.48	2.66	60.2
RP	RP:XB2	End	39.00	10.19	0.54	-0.18	-47.22	-3.24	0.6	-19.02	-3.27	-0.05	0.18	3.74	3.92	87.9
RP	RP:XB2	Origin	39.00	10.19	0.54	-0.18	-50.89	-3.23	0.6	-26.63	4.14	0.01	0.26	4.03	4.28	96.1
RP	#RP:17	End	44.00	8.75	0.35	-0.15	-30.18	-3.21	0.6	-26.63	4.14	0.01	0.21	1.83	2.04	44.8
RP	#RP:17	Origin	44.00	8.75	0.35	-0.15	-30.18	-3.20	0.6	-26.99	4.43	0.03	0.21	1.83	2.04	44.8
RP	#RP:18	End	49.00	6.83	0.22	-0.12	-8.05	-3.06	0.6	-26.99	4.43	0.03	0.20	0.46	0.66	14.1
RP	#RP:18	Origin	49.00	6.83	0.22	-0.12	-8.06	-3.06	0.6	-27.40	4.55	0.05	0.20	0.46	0.66	14.1
RP	#RP:19	End	54.00	4.79	0.12	-0.08	14.68	-2.83	0.6	-27.40	4.55	0.05	0.19	0.71	0.89	18.7
RP	#RP:19	Origin	54.00	4.79	0.12	-0.08	14.68	-2.83	0.6	-27.88	4.54	0.06	0.19	0.71	0.90	18.8
RP	#RP:20	End	59.00	2.91	0.06	-0.04	37.40	-2.52	0.6	-27.88	4.54	0.06	0.17	1.58	1.76	36.0
RP	#RP:20	Origin	59.00	2.91	0.06	-0.04	37.40	-2.52	0.6	-28.41	4.44	0.07	0.18	1.58	1.76	36.1
RP	#RP:21	End	64.00	1.40	0.02	-0.02	59.62	-2.15	0.6	-28.41	4.44	0.07	0.17	2.25	2.42	48.6
RP	#RP:21	Origin	64.00	1.40	0.02	-0.02	59.62	-2.16	0.6	-28.98	4.27	0.08	0.17	2.25	2.42	48.6
RP	#RP:22	End	69.00	0.40	0.01	-0.01	80.99	-1.75	0.6	-28.98	4.27	0.08	0.16	2.75	2.91	57.2
RP	#RP:22	Origin	69.00	0.40	0.01	-0.01	80.99	-1.75	0.6	-29.45	4.10	0.09	0.16	2.75	2.91	57.2
RP	#RP:23	End	71.75	0.10	0.00	-0.00	92.27	-1.51	0.6	-29.45	4.10	0.09	0.15	2.96	3.11	60.6
RP	#RP:23	Origin	71.75	0.10	0.00	-0.00	92.27	-1.52	0.6	-29.80	3.97	0.09	0.16	2.96	3.12	60.6
RP	RP:g	End	74.50	0.00	0.00	0.00	103.18	-1.27	0.6	-29.80	3.97	0.09	0.15	3.13	3.28	63.1

DESIGN PARAMETERS:	TRANSMISSION POLE DESIGN EXAMPLE 2
85` C2 SOUTHERN PINE POLES WEIGHT OF WOOD 60 PCF MOR = 8,000 PSI Strength Factor = 0.671 MOE = 2,130,000 PSI	ASCE WOOD POLE STRUCTURE MANUAL OF PRACTICE

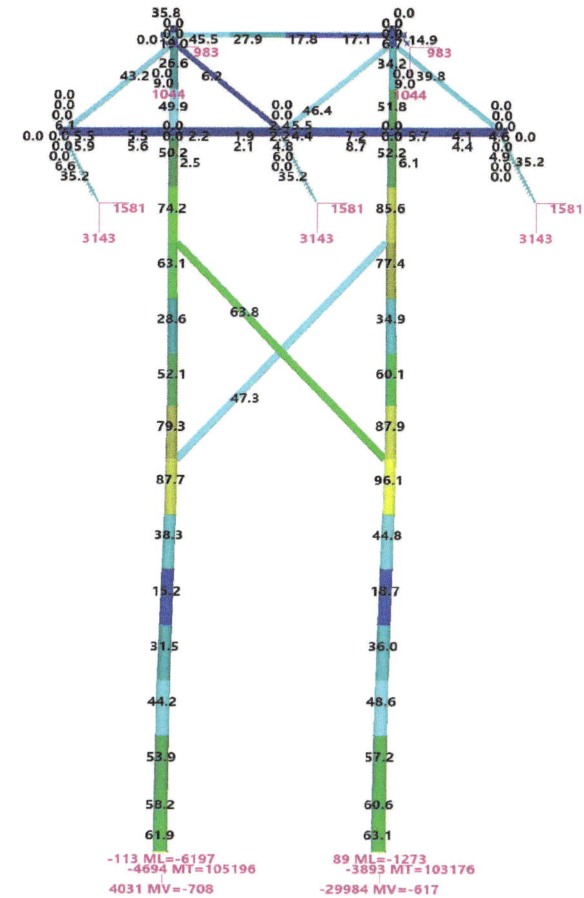

DESIGN PARAMETERS:	TRANSMISSION POLE DESIGN EXAMPLE 2
85` C2 SOUTHERN PINE POLES WEIGHT OF WOOD 60 PCF MOR = 8,000 PSI Strength Factor = 0.671 MOE = 2,130,000 PSI	ASCE WOOD POLE STRUCTURE MANUAL OF PRACTICE

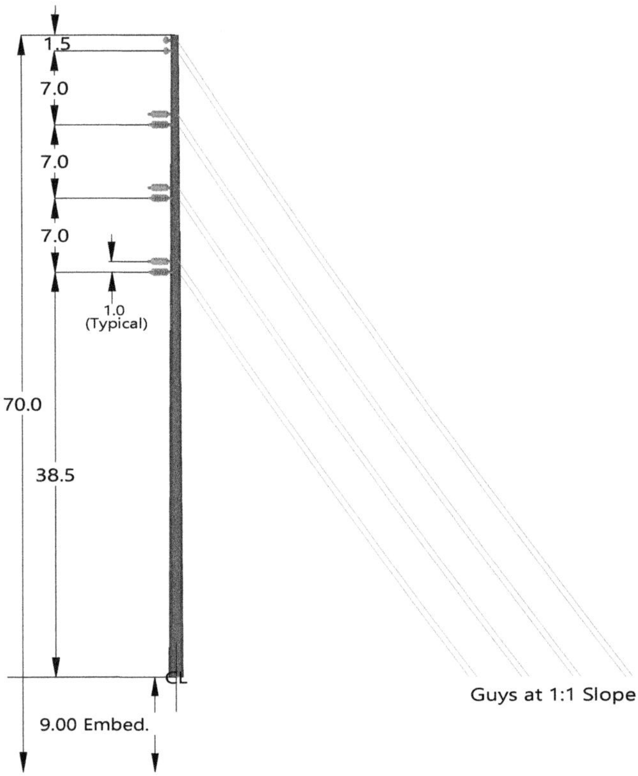

DESIGN PARAMETERS: 70` C2 WESTERN RED CEDAR WEIGHT OF WOOD 50 PCF MOR = 6,000 PSI Strength Factor = 0.671 MOE = 1,4300,000 PSI	TRANSMISSION POLE DESIGN EXAMPLE 3 ASCE WOOD POLE STRUCTURE MANUAL OF PRACTICE

Loading Tree					
Load Case	Joint Label	Vert.	Trans.	Long.	Trans. Wind (psf)
		Load (lbs)			
ASCE Design Example 3	SA	243	-4985	-4709	-10
	LA	560	-5478	-5045	-10
	CA	560	-5478	-5045	-10
	RA	560	-5478	-5045	-10
	SB	243	-4984	4709	-10
	LB	560	-5477	5046	-10
	CB	560	-5477	5046	-10
	RB	560	-5477	5046	-10

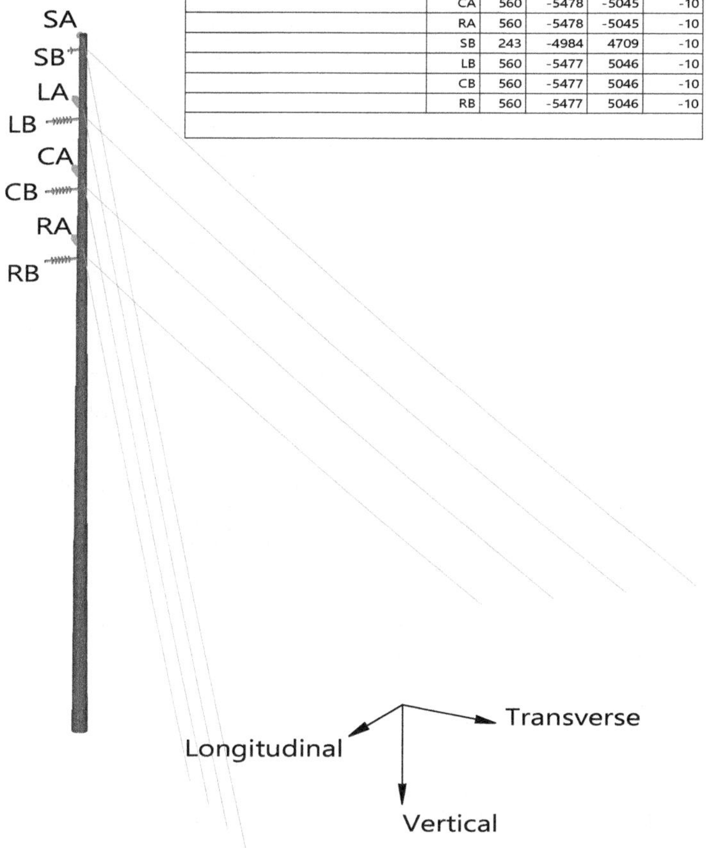

DESIGN PARAMETERS:
70` C2 WESTERN
RED CEDAR
WEIGHT OF WOOD 50 PCF
MOR = 6,000 PSI
Strength Factor = 0.671
MOE = 1,4300,000 PSI

TRANSMISSION POLE
DESIGN EXAMPLE 3

ASCE WOOD POLE STRUCTURE
MANUAL OF PRACTICE

Detailed Wood Pole Usages for Load Case "ASCE Design Example 3"

Element Label	Joint Label	Joint Position	Rel. Dist. (ft)	Trans. Defl. (in)	Long. Defl. (in)	Vert. Defl. (in)	Trans. Mom. (ft-k)	Long. Mom. (ft-k)	Tors. Mom. (ft-k)	Axial Force (kips)	Tran. Shear (kips)	Long. Shear (kips)	P/A (ksi)	M/S. (ksi)	Res. (ksi)	Max. Usage %
P	P:t	Origin	0.00	-7.43	-0.02	-0.27	0.00	-0.00	-0.0	-0.01	-0.00	-0.00	0.00	0. 00	0.00	0.0
P	P:SA	End	0.50	-7.43	-0.02	-0.27	-0.00	-0.00	-0.0	-0.01	-0.00	-0.00	0.00	0.00	0.00	0.0
P	P:SA	Origin	0.50	-7.43	-0.02	-0.27	1.55	-1.55	0.1	-7.07	-0.14	-0.09	0.16	0. 63	0.79	23.0
P	P:SB	End	1.50	-7.42	-0.03	-0.27	1.41	-1.64	0.1	-7.07	-0.14	-0.09	0.15	0.49	0.64	18.7
P	P:SB	Origin	1.50	-7.42	-0.03	-0.27	3.01	-0.04	-0.0	-14.26	-0.26	0.06	0.31	0.74	1.05	30.6
P	#P:0	End	4.50	-7.32	-0.02	-0.27	2.23	0.15	-0.0	-14.26	-0.26	0.06	0.25	0.43	0.68	19.8
P	#P:0	Origin	4.50	-7.32	-0.02	-0.27	2.23	0.15	-0.0	-14.35	-0.32	0.06	0.25	0. 43	0.68	19.9
P	P:CA	End	7.50	-7.13	-0.02	-0.26	1.29	0.33	-0.0	-14.35	-0.32	0.06	0.25	0.23	0.49	14.2
P	P:CA	Origin	7.50	-7.13	-0.02	-0.26	3.17	-1.55	0.1	-22.58	-0.44	-0.27	0.40	0 .68	1.09	31.7
P	P:CB	End	8.50	-7.05	-0.02	-0.26	2.73	-1.81	0.1	-22.58	-0.44	-0.27	0.39	0.53	0.92	26.8
P	P:CB	Origin	8.50	-7.05	-0.02	-0.26	4.65	0.10	-0.0	-30.85	-0.58	0.09	0.53	0. 81	1.34	39.1
P	#P:1	End	11.50	-6.74	-0.01	-0.24	2.91	0.37	-0.0	-30.85	-0.58	0.09	0.43	0.41	0.84	24.6
P	#P:1	Origin	11.50	-6.74	-0.01	-0.24	2.91	0.37	-0.0	-30.96	-0.67	0.08	0.43	0.41	0.84	24.6
P	P:LA	End	14.50	-6.36	-0.02	-0.23	0.90	0.61	-0.0	-30.96	-0.67	0.08	0.44	0.14	0.58	16.9
P	P:LA	Origin	14.50	-6.36	-0.02	-0.23	3.10	-1.59	0.1	-39.57	-0.58	-0.51	0.56	0.49	1.06	30.8
P	P:LB	End	15.50	-6.22	-0.02	-0.23	2.52	-2.10	0.1	-39.57	-0.58	-0.51	0.55	0.3 9	0.94	27.4
P	P:LB	Origin	15.50	-6.22	-0.02	-0.23	4.75	0.13	-0.0	-48.17	-0.54	0.07	0.67	0.60	1.27	37.2
P	#P:2	End	18.50	-5.75	-0.01	-0.21	3.11	0.34	-0.0	-48.17	-0.54	0.07	0.55	0.33	0.88	25.7
P	#P:2	Origin	18.50	-5.75	-0.01	-0.21	3.11	0.34	-0.0	-48.31	-0.65	0.06	0.55	0.33	0.88	25.8
P	P:RA	End	21.50	-5.23	-0.01	-0.19	1.17	0.53	-0.0	-48.31	-0.65	0.06	0.57	0.12	0.69	20.1
P	P:RA	Origin	21.50	-5.23	-0.01	-0.19	3.56	-1.86	0.1	-56.84	-0.67	-0.45	0.67	0.42	1.09	31.8
P	P:RB	End	22.50	-5.05	-0.01	-0.18	2.89	-2.32	0.1	-56.84	-0.67	-0.45	0.65	0.3 3	0.98	28.7
P	P:RB	Origin	22.50	-5.05	-0.01	-0.18	5.28	0.07	-0.0	-65.33	-0.82	0.00	0.75	0.51	1.25	36.6
P	#P:3	End	27.50	-4.05	-0.01	-0.15	1.18	0.09	-0.0	-65.33	-0.82	0.00	0.60	0.09	0.69	20.1
P	#P:3	Origin	27.50	-4.05	-0.01	-0.15	1.18	0.09	-0.0	-65.62	-0.92	0.00	0.60	0.09	0.69	20.1
P	#P:4	End	32.50	-3.02	-0.00	-0.12	-3.42	0.10	-0.0	-65.62	-0.92	0.00	0.54	0.2 1	0.75	21.7
P	#P:4	Origin	32.50	-3.02	-0.00	-0.12	-3.42	0.10	-0.0	-65.94	-0.89	-0.00	0.5 4	0.21	0.75	21.8
P	#P:5	End	37.50	-2.06	0.00	-0.09	-7.84	0.08	-0.0	-65.94	-0.89	-0.00	0.48	0.4 2	0.90	25.3
P	#P:5	Origin	37.50	-2.06	0.00	-0.09	-7.84	0.08	-0.0	-66.30	-0.77	-0.01	0.49	0.42	0.90	25.4
P	#P:6	End	42.50	-1.27	0.00	-0.06	-11.71	0.06	-0.0	-66.30	-0.77	-0.01	0.44	0. 54	0.97	26.6
P	#P:6	Origin	42.50	-1.27	0.00	-0.06	-11.71	0.06	-0.0	-66.69	-0.63	-0.01	0.4 4	0.54	0.98	26.7
P	#P:7	End	47.50	-0.66	0.00	-0.04	-14.84	0.02	-0.0	-66.69	-0.63	-0.01	0.40	0. 59	0.99	26.2
P	#P:7	Origin	47.50	-0.66	0.00	-0.04	-14.84	0.02	-0.0	-67.13	-0.47	-0.01	0.4 0	0.59	0.99	26.3
P	#P:8	End	52.50	-0.25	0.00	-0.02	-17.17	-0.01	-0.0	-67.13	-0.47	-0.01	0.37	0 .59	0.96	24.8
P	#P:8	Origin	52.50	-0.25	0.00	-0.02	-17.17	-0.01	-0.0	-67.57	-0.33	-0.01	0. 37	0.59	0.96	24.8
P	#P:9	End	56.75	-0.06	0.00	-0.01	-18.55	-0.05	-0.0	-67.57	-0.33	-0.01	0.34	0 .57	0.91	23.1
P	#P:9	Origin	56.75	-0.06	0.00	-0.01	-18.55	-0.05	-0.0	-68.01	-0.21	-0.01	0. 34	0.57	0.91	23.1
P	P:g	End	61.00	0.00	0.00	0.00	-19.44	-0.09	-0.0	-68.01	-0.21	-0.01	0.32	0.53	0.85	21.2

Summary of Guy Tensions and Usages for Load Case "ASCE Design Example 3"

Guy Label	Max. Tension (kips)	Allowable Tension (kips)	Factored Allowable (kips)	Usage %
SAG	9.63	13.05	11.75	82.01
SBG	9.76	13.05	11.75	83.10
LAG	11.31	13.05	11.75	96.28
LBG	11.28	13.05	11.75	96.03
CAG	10.79	13.05	11.74	91.87
CBG	10.83	13.05	11.74	92.24
RAG	11.17	13.05	11.75	95.07
RBG	11.03	13.05	11.74	93.95

DESIGN PARAMETERS:	TRANSMISSION POLE DESIGN EXAMPLE 3
70` C2 WESTERN RED CEDAR WEIGHT OF WOOD 50 PCF MOR = 6,000 PSI Strength Factor = 0.671 MOE = 1,4300,000 PSI	ASCE WOOD POLE STRUCTURE MANUAL OF PRACTICE

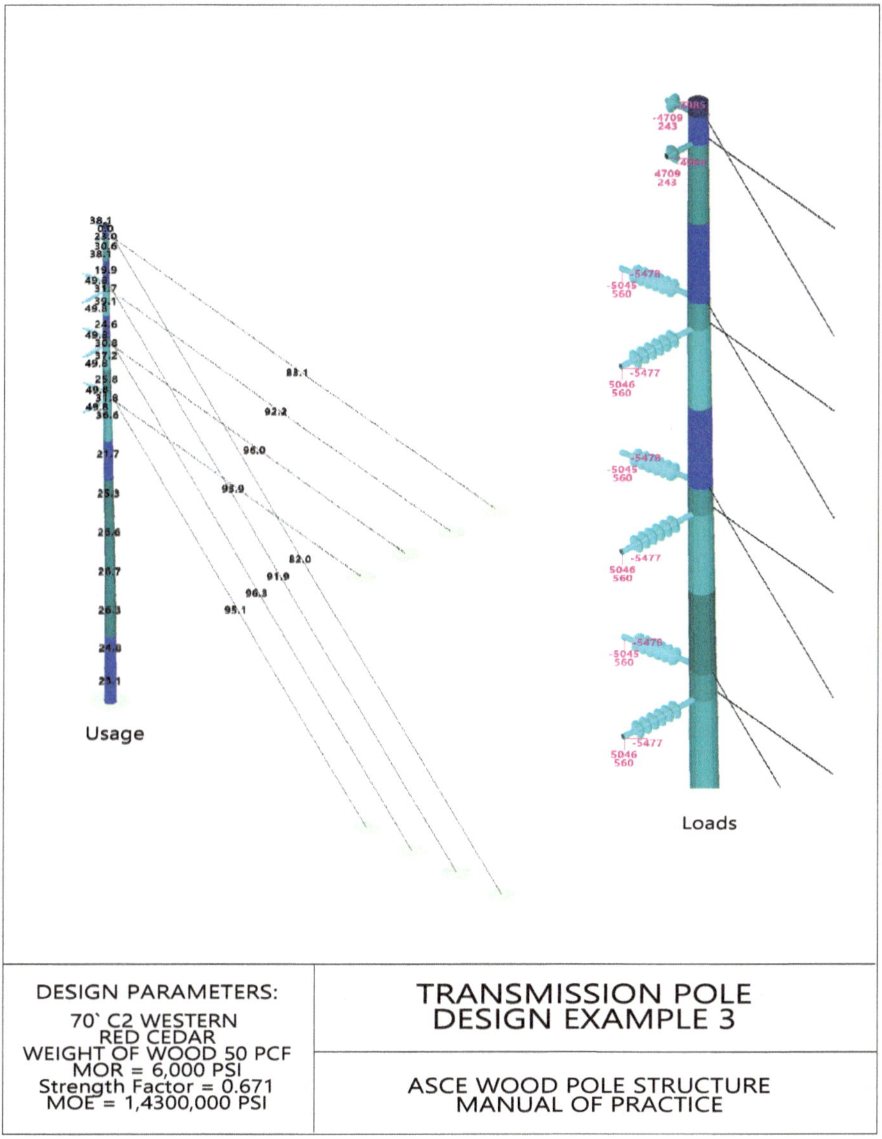

DESIGN PARAMETERS:
70\` C2 WESTERN
RED CEDAR
WEIGHT OF WOOD 50 PCF
MOR = 6,000 PSI
Strength Factor = 0.671
MOE = 1,4300,000 PSI

TRANSMISSION POLE
DESIGN EXAMPLE 3

ASCE WOOD POLE STRUCTURE
MANUAL OF PRACTICE

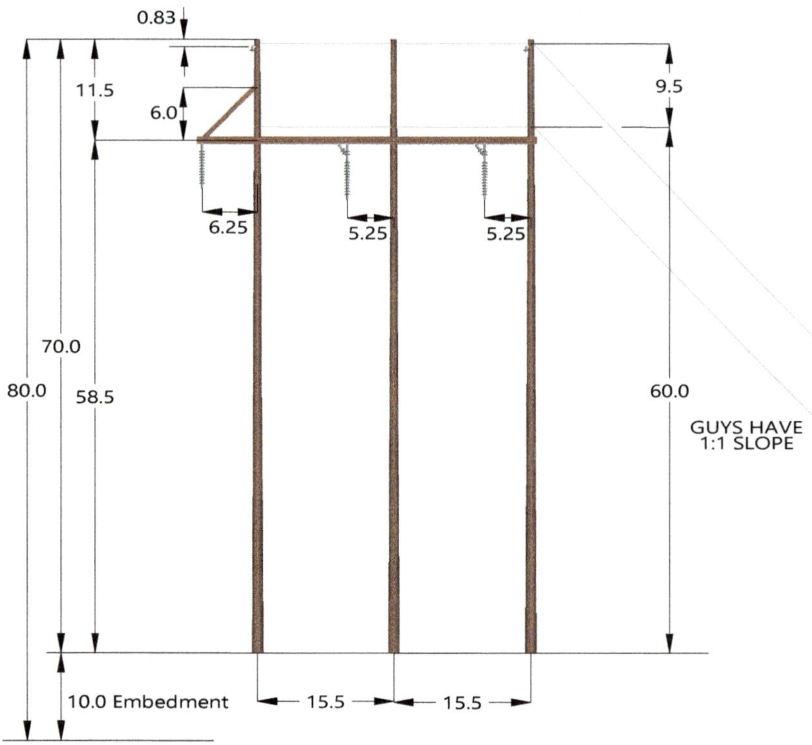

DESIGN PARAMETERS:	TRANSMISSION POLE DESIGN EXAMPLE 4
80` C3 DOUGLAS FIR POLES WEIGHT OF WOOD 60 PCF MOR = 8,000 PSI Strength Factor = 0.671 MOE = 2,380,000 PSI	ASCE WOOD POLE STRUCTURE MANUAL OF PRACTICE

Loading Tree

Load Case	Joint Label	Vert.	Trans.	Long.	Trans. Wind (psf)
		Load (lbs)			
ASCE Design Example 4,I NA-	S1	986	-1899	0	-10
	S2	986	-1899	0	-10
	L	2726	-3349	0	-10
	C	2726	-3349	0	-10
	R	2726	-3349	0	-10

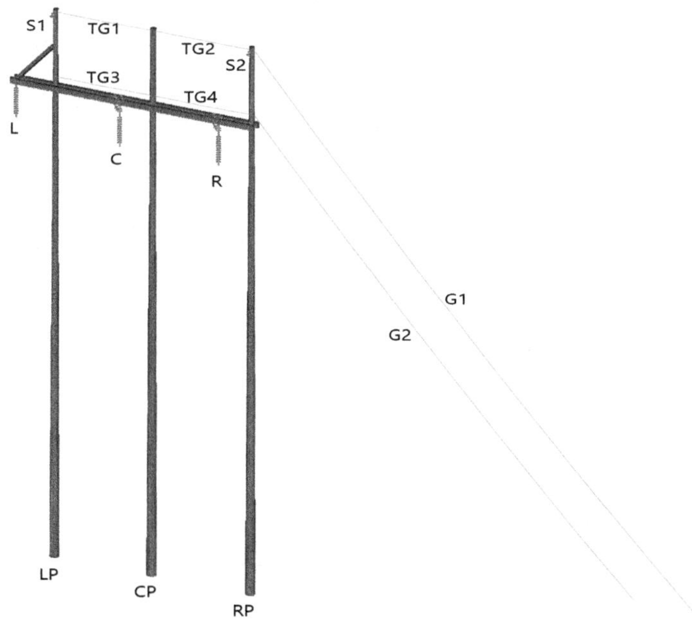

DESIGN PARAMETERS:
80` C3
DOUGLAS FIR POLES
WEIGHT OF WOOD 60 PCF
MOR = 8,000 PSI
Strength Factor = 0.671
MOE = 2,380,000 PSI

TRANSMISSION POLE DESIGN EXAMPLE 4

ASCE WOOD POLE STRUCTURE MANUAL OF PRACTICE

Detailed Wood Pole Usages for Load Case "ASCE Design Example 4,I NA-" (1 of 3)

Element Label	Joint Label	Joint Position	Rel. Dist. (ft)	Trans. Defl. (in)	Long. Defl. (in)	Vert. Defl. (in)	Trans. Mom. (ft-k)	Long. Mom. (ft-k)	Tors. Mom. (ft-k)	Axial Force (kips)	Tran. Shear (kips)	Long. Shear (kips)	P/A (ksi)	M/S. (ksi)	Res. (ksi)	Max. Usage %
LP	LP:t	Origin	0.00	-5.49	0.33	-0.06	0.00	-0.00	-0.0	-0.01	-0.00	-0.00	0.00	0.00	0.00	0.0
LP	LP:G1	End	0.50	-5.57	0.33	-0.06	-0.00	-0.00	-0.0	-0.01	-0.00	-0.00	0.00	0.00	0.00	0.0
LP	LP:G1	Origin	0.50	-5.57	0.33	-0.06	-0.00	-0.00	-0.0	0.03	3.57	-0.00	-0.00	0.00	0.00	0.0
LP	LP:S	End	0.83	-5.62	0.33	-0.06	1.19	-0.00	-0.0	0.03	3.57	-0.00	-0.00	0.44	0.44	9.6
LP	LP:S	Origin	0.83	-5.62	0.33	-0.06	0.64	-0.00	0.0	-1.06	1.67	-0.00	0.03	0.24	0.26	5.8
LP	LP:KB1	End	5.50	-6.20	0.29	-0.05	8.41	-0.00	0.0	-1.06	1.67	-0.00	0.03	2.62	2.65	58.1
LP	LP:KB1	Origin	5.50	-6.20	0.29	-0.05	7.59	-0.00	0.0	-3.72	-0.83	-0.00	0.09	2.37	2.46	53.8
LP	LP:G2	End	10.00	-6.28	0.25	-0.05	3.86	-0.01	0.0	-3.72	-0.83	-0.00	0.08	0.97	1.05	22.9
LP	LP:G2	Origin	10.00	-6.28	0.25	-0.05	3.86	-0.01	0.0	-3.82	0.65	-0.00	0.08	0.97	1.05	23.0
LP	LP:ARM	End	11.50	-6.24	0.24	-0.05	4.84	-0.01	0.0	-3.82	0.65	-0.00	0.08	0.96	1.04	22.8
LP	LP:ARM	Origin	11.50	-6.24	0.24	-0.05	4.84	-0.01	0.0	-5.29	-0.14	-0.01	0.11	0.96	1.07	23.5
LP	#LP:0	End	16.50	-5.90	0.20	-0.05	4.13	-0.04	0.0	-5.29	-0.14	-0.01	0.08	0.69	0.77	16.9
LP	#LP:0	Origin	16.50	-5.90	0.20	-0.05	4.13	-0.04	0.0	-5.49	-0.20	-0.01	0.09	0.69	0.77	16.9
LP	#LP:1	End	21.50	-5.32	0.17	-0.04	3.12	-0.07	0.0	-5.49	-0.20	-0.01	0.08	0.44	0.52	11.3
LP	#LP:1	Origin	21.50	-5.32	0.17	-0.04	3.12	-0.07	0.0	-5.71	-0.26	-0.01	0.08	0.44	0.52	11.4
LP	#LP:2	End	26.50	-4.61	0.13	-0.04	1.84	-0.10	0.0	-5.71	-0.26	-0.01	0.07	0.22	0.29	6.4
LP	#LP:2	Origin	26.50	-4.61	0.13	-0.04	1.84	-0.10	0.0	-5.96	-0.31	-0.01	0.07	0.22	0.30	6.5
LP	#LP:3	End	31.50	-3.83	0.10	-0.03	0.29	-0.13	0.0	-5.96	-0.31	-0.01	0.07	0.03	0.10	2.2
LP	#LP:3	Origin	31.50	-3.83	0.10	-0.03	0.29	-0.13	0.0	-6.23	-0.36	-0.01	0.07	0.03	0.10	2.3
LP	#LP:4	End	36.50	-3.04	0.08	-0.02	-1.49	-0.16	0.0	-6.23	-0.36	-0.01	0.06	0.13	0.20	4.3
LP	#LP:4	Origin	36.50	-3.04	0.08	-0.02	-1.49	-0.16	0.0	-6.53	-0.40	-0.01	0.07	0.13	0.20	4.4
LP	#LP:5	End	41.50	-2.29	0.06	-0.02	-3.51	-0.19	0.0	-6.53	-0.40	-0.01	0.06	0.27	0.34	7.1
LP	#LP:5	Origin	41.50	-2.29	0.06	-0.02	-3.51	-0.19	0.0	-6.86	-0.45	-0.01	0.06	0.27	0.34	7.2
LP	#LP:6	End	46.50	-1.61	0.04	-0.01	-5.75	-0.22	0.0	-6.86	-0.45	-0.01	0.06	0.39	0.45	9.4
LP	#LP:6	Origin	46.50	-1.61	0.04	-0.01	-5.75	-0.22	0.0	-7.23	-0.49	-0.01	0.06	0.39	0.46	9.5
LP	#LP:7	End	51.50	-1.02	0.02	-0.01	-8.20	-0.25	0.0	-7.23	-0.49	-0.01	0.06	0.50	0.55	11.2
LP	#LP:7	Origin	51.50	-1.02	0.02	-0.01	-8.20	-0.25	0.0	-7.62	-0.53	-0.01	0.06	0.50	0.56	11.3
LP	#LP:8	End	56.50	-0.56	0.01	-0.01	-10.86	-0.29	0.0	-7.62	-0.53	-0.01	0.06	0.58	0.64	12.6
LP	#LP:8	Origin	56.50	-0.56	0.01	-0.01	-10.86	-0.29	0.0	-8.05	-0.57	-0.01	0.06	0.58	0.64	12.7
LP	#LP:9	End	61.50	-0.22	0.00	-0.00	-13.72	-0.32	0.0	-8.05	-0.57	-0.01	0.05	0.66	0.71	13.8
LP	#LP:9	Origin	61.50	-0.22	0.00	-0.00	-13.72	-0.32	0.0	-8.47	-0.61	-0.01	0.06	0.66	0.71	13.8
LP	#LP:10	End	65.75	-0.06	0.00	-0.00	-16.31	-0.35	0.0	-8.47	-0.61	-0.01	0.05	0.71	0.76	14.5
LP	#LP:10	Origin	65.75	-0.06	0.00	-0.00	-16.31	-0.35	0.0	-8.89	-0.64	-0.01	0.06	0.71	0.77	14.5
LP	LP:g	End	70.00	0.00	0.00	0.00	-19.03	-0.39	0.0	-8.89	-0.64	-0.01	0.05	0.76	0.81	15.1

Detailed Wood Pole Usages for Load Case "ASCE Design Example 4,I NA-" (2 of 3)

Element Label	Joint Label	Joint Position	Rel. Dist. (ft)	Trans. Defl. (in)	Long. Defl. (in)	Vert. Defl. (in)	Trans. Mom. (ft-k)	Long. Mom. (ft-k)	Tors. Mom. (ft-k)	Axial Force (kips)	Tran. Shear (kips)	Long. Shear (kips)	P/A (ksi)	M/S. (ksi)	Res. (ksi)	Max. Usage %
CP	CP:t	Origin	0.00	-5.30	0.33	-0.06	0.00	-0.00	-0.0	-0.01	-0.00	-0.00	0.00	0.00	0.00	0.0
CP	CP:G1	End	0.50	-5.36	0.33	-0.06	-0.00	-0.00	-0.0	-0.01	-0.00	-0.00	0.00	0.00	0.00	0.0
CP	CP:G1	Origin	0.50	-5.36	0.33	-0.06	-0.00	-0.00	-0.0	-0.02	0.54	-0.00	0.00	0.00	0.00	0.0
CP	CP:S	End	0.83	-5.40	0.32	-0.06	0.18	-0.00	-0.0	-0.02	0.54	-0.00	0.00	0.05	0.05	1.2
CP	CP:S	Origin	0.83	-5.40	0.32	-0.06	0.18	-0.00	0.0	-0.09	0.53	-0.00	0.00	0.05	0.06	1.2
CP	CP:KB1	End	5.50	-5.91	0.29	-0.05	2.64	-0.00	0.0	-0.09	0.53	-0.00	0.00	0.66	0.66	14.4
CP	CP:KB1	Origin	5.50	-5.91	0.29	-0.05	2.64	-0.00	0.0	-0.23	0.50	-0.00	0.00	0.66	0.66	14.5
CP	CP:G2	End	10.00	-6.20	0.25	-0.05	4.87	-0.01	0.0	-0.23	0.50	-0.00	0.00	1.22	1.23	26.9
CP	CP:G2	Origin	10.00	-6.20	0.25	-0.05	4.87	-0.01	-0.0	-0.34	1.27	-0.00	0.01	1.22	1.23	27.0
CP	CP:ARM	End	11.50	-6.24	0.24	-0.05	6.77	-0.01	-0.0	-0.34	1.27	-0.00	0.01	1.35	1.35	29.7
CP	CP:ARM	Origin	11.50	-6.24	0.24	-0.05	6.77	-0.02	0.0	-5.02	-0.19	-0.00	0.10	1.35	1.45	31.8
CP	#CP:11	End	16.50	-6.08	0.20	-0.05	5.82	-0.05	0.0	-5.02	-0.19	-0.00	0.08	0.97	1.05	22.9
CP	#CP:11	Origin	16.50	-6.08	0.20	-0.05	5.82	-0.05	0.0	-5.22	-0.26	-0.01	0.08	0.97	1.05	23.0
CP	#CP:12	End	21.50	-5.59	0.16	-0.05	4.53	-0.07	0.0	-5.22	-0.26	-0.01	0.07	0.64	0.71	15.6
CP	#CP:12	Origin	21.50	-5.59	0.16	-0.05	4.53	-0.07	0.0	-5.44	-0.32	-0.01	0.08	0.64	0.71	15.7
CP	#CP:13	End	26.50	-4.91	0.13	-0.04	2.95	-0.10	0.0	-5.44	-0.32	-0.01	0.07	0.35	0.42	9.3
CP	#CP:13	Origin	26.50	-4.91	0.13	-0.04	2.95	-0.10	0.0	-5.69	-0.37	-0.01	0.07	0.35	0.43	9.3
CP	#CP:14	End	31.50	-4.11	0.10	-0.03	1.09	-0.13	0.0	-5.69	-0.37	-0.01	0.06	0.11	0.18	3.9
CP	#CP:14	Origin	31.50	-4.11	0.10	-0.03	1.09	-0.13	0.0	-5.96	-0.42	-0.01	0.07	0.11	0.18	4.0
CP	#CP:15	End	36.50	-3.28	0.08	-0.03	-1.03	-0.15	0.0	-5.96	-0.42	-0.01	0.06	0.09	0.15	3.4
CP	#CP:15	Origin	36.50	-3.28	0.08	-0.03	-1.03	-0.15	0.0	-6.26	-0.47	-0.01	0.06	0.09	0.16	3.4
CP	#CP:16	End	41.50	-2.48	0.05	-0.02	-3.38	-0.18	0.0	-6.26	-0.47	-0.01	0.06	0.26	0.32	6.9
CP	#CP:16	Origin	41.50	-2.48	0.05	-0.02	-3.38	-0.18	0.0	-6.59	-0.52	-0.01	0.06	0.26	0.33	6.9
CP	#CP:17	End	46.50	-1.75	0.04	-0.01	-5.96	-0.22	0.0	-6.59	-0.52	-0.01	0.06	0.41	0.47	9.7
CP	#CP:17	Origin	46.50	-1.75	0.04	-0.01	-5.96	-0.22	0.0	-6.96	-0.56	-0.01	0.06	0.41	0.47	9.7
CP	#CP:18	End	51.50	-1.11	0.02	-0.01	-8.76	-0.25	0.0	-6.96	-0.56	-0.01	0.06	0.53	0.59	11.9
CP	#CP:18	Origin	51.50	-1.11	0.02	-0.01	-8.76	-0.25	0.0	-7.35	-0.60	-0.01	0.06	0.53	0.59	11.9
CP	#CP:19	End	56.50	-0.61	0.01	-0.01	-11.76	-0.28	0.0	-7.35	-0.60	-0.01	0.05	0.63	0.69	13.6
CP	#CP:19	Origin	56.50	-0.61	0.01	-0.01	-11.76	-0.28	0.0	-7.78	-0.64	-0.01	0.06	0.63	0.69	13.6
CP	#CP:20	End	61.50	-0.25	0.00	-0.00	-14.96	-0.32	0.0	-7.78	-0.64	-0.01	0.05	0.72	0.77	14.9
CP	#CP:20	Origin	61.50	-0.25	0.00	-0.00	-14.96	-0.32	0.0	-8.20	-0.68	-0.01	0.06	0.72	0.77	14.9
CP	#CP:21	End	65.75	-0.06	0.00	-0.00	-17.83	-0.35	0.0	-8.20	-0.68	-0.01	0.05	0.78	0.83	15.7
CP	#CP:21	Origin	65.75	-0.06	0.00	-0.00	-17.83	-0.35	0.0	-8.62	-0.71	-0.01	0.06	0.78	0.83	15.8
CP	CP:g	End	70.00	0.00	0.00	0.00	-20.83	-0.38	0.0	-8.62	-0.71	-0.01	0.05	0.83	0.88	16.4

DESIGN PARAMETERS:	TRANSMISSION POLE DESIGN EXAMPLE 4
80` C3 DOUGLAS FIR POLES WEIGHT OF WOOD 60 PCF MOR = 8,000 PSI Strength Factor = 0.671 MOE = 2,380,000 PSI	ASCE WOOD POLE STRUCTURE MANUAL OF PRACTICE

Detailed Wood Pole Usages for Load Case "ASCE Design Example 4,I NA-" (3 of 3)

Element Label	Joint Label	Joint Position	Rel. Dist. (ft)	Trans. Defl. (in)	Long. Defl. (in)	Vert. Defl. (in)	Trans. Mom. (ft-k)	Long. Mom. (ft-k)	Tors. Mom. (ft-k)	Axial Force (kips)	Tran. Shear (kips)	Long. Shear (kips)	P/A (ksi)	M/S. (ksi)	Res. (ksi)	Max. Usage %
RP	RP:t	Origin	0.00	-5.06	0.34	-0.11	0.00	-0.00	-0.0	-0.01	-0.00	-0.00	0.00	0.00	0.00	0.0
RP	RP:G1	End	0.50	-5.12	0.34	-0.11	-0.00	-0.00	-0.0	-0.01	-0.00	-0.00	0.00	0.00	0.00	0.0
RP	RP:G1	Origin	0.50	-5.12	0.34	-0.11	1.70	-0.00	-0.0	-5.60	1.52	-0.00	0.15	0.64	0.79	17.3
RP	RP:S	End	0.83	-5.15	0.33	-0.11	2.21	-0.00	-0.0	-5.60	1.52	-0.00	0.15	0.82	0.97	21.2
RP	RP:S	Origin	0.83	-5.15	0.33	-0.11	1.66	-0.00	-0.0	-6.67	-0.39	-0.00	0.18	0.61	0.79	17.4
RP	RP:KB1	End	5.50	-5.62	0.29	-0.10	-0.16	-0.01	-0.0	-6.67	-0.39	-0.00	0.14	0.04	0.18	3.9
RP	RP:KB1	Origin	5.50	-5.62	0.29	-0.10	-0.16	-0.01	-0.0	-6.81	-0.42	-0.00	0.14	0.04	0.18	3.9
RP	RP:G2	End	10.00	-6.07	0.25	-0.10	-2.04	-0.03	-0.0	-6.81	-0.42	-0.00	0.14	0.51	0.65	14.3
RP	RP:G2	Origin	10.00	-6.07	0.25	-0.10	0.79	-0.03	-0.0	-15.07	5.58	-0.01	0.31	0.20	0.51	11.2
RP	RP:ARM	End	11.50	-6.23	0.24	-0.10	9.16	-0.04	-0.0	-15.07	5.58	-0.01	0.31	1.82	2.13	46.7
RP	RP:ARM	Origin	11.50	-6.23	0.24	-0.10	9.16	-0.03	0.0	-17.57	-0.09	-0.01	0.36	1.82	2.18	47.8
RP	#RP:22	End	16.50	-6.40	0.19	-0.09	8.70	-0.06	0.0	-17.57	-0.09	-0.01	0.27	1.45	1.72	37.7
RP	#RP:22	Origin	16.50	-6.40	0.19	-0.09	8.70	-0.06	0.0	-17.77	-0.27	-0.01	0.28	1.45	1.72	37.8
RP	#RP:23	End	21.50	-6.09	0.15	-0.08	7.34	-0.10	0.0	-17.77	-0.27	-0.01	0.25	1.03	1.28	28.1
RP	#RP:23	Origin	21.50	-6.09	0.15	-0.08	7.34	-0.10	0.0	-17.99	-0.41	-0.01	0.25	1.03	1.28	28.2
RP	#RP:24	End	26.50	-5.46	0.12	-0.07	5.29	-0.13	0.0	-17.99	-0.41	-0.01	0.23	0.64	0.86	18.9
RP	#RP:24	Origin	26.50	-5.46	0.12	-0.07	5.29	-0.13	0.0	-18.24	-0.51	-0.01	0.23	0.64	0.86	19.0
RP	#RP:25	End	31.50	-4.63	0.09	-0.06	2.72	-0.16	0.0	-18.24	-0.51	-0.01	0.21	0.28	0.49	10.7
RP	#RP:25	Origin	31.50	-4.63	0.09	-0.06	2.72	-0.16	0.0	-18.51	-0.59	-0.00	0.21	0.28	0.49	10.8
RP	#RP:26	End	36.50	-3.72	0.07	-0.05	-0.21	-0.19	0.0	-18.51	-0.59	-0.00	0.19	0.03	0.22	4.7
RP	#RP:26	Origin	36.50	-3.72	0.07	-0.05	-0.21	-0.19	0.0	-18.81	-0.64	-0.00	0.19	0.03	0.22	4.8
RP	#RP:27	End	41.50	-2.82	0.05	-0.04	-3.39	-0.21	0.0	-18.81	-0.64	-0.00	0.18	0.27	0.44	9.4
RP	#RP:27	Origin	41.50	-2.82	0.05	-0.04	-3.39	-0.21	0.0	-19.15	-0.67	-0.00	0.18	0.27	0.45	9.5
RP	#RP:28	End	46.50	-1.98	0.03	-0.03	-6.73	-0.22	0.0	-19.15	-0.67	-0.00	0.17	0.46	0.63	13.0
RP	#RP:28	Origin	46.50	-1.98	0.03	-0.03	-6.73	-0.22	0.0	-19.51	-0.69	-0.00	0.17	0.46	0.63	13.1
RP	#RP:29	End	51.50	-1.26	0.02	-0.02	-10.16	-0.24	0.0	-19.51	-0.69	-0.00	0.16	0.62	0.77	15.6
RP	#RP:29	Origin	51.50	-1.26	0.02	-0.02	-10.16	-0.24	0.0	-19.90	-0.69	-0.00	0.16	0.62	0.77	15.7
RP	#RP:30	End	56.50	-0.68	0.01	-0.01	-13.64	-0.25	0.0	-19.90	-0.69	-0.00	0.15	0.73	0.88	17.4
RP	#RP:30	Origin	56.50	-0.68	0.01	-0.01	-13.64	-0.25	0.0	-20.33	-0.70	-0.00	0.15	0.73	0.88	17.4
RP	#RP:31	End	61.50	-0.27	0.00	-0.01	-17.12	-0.26	0.0	-20.33	-0.70	-0.00	0.14	0.82	0.96	18.5
RP	#RP:31	Origin	61.50	-0.27	0.00	-0.01	-17.12	-0.26	0.0	-20.75	-0.69	-0.00	0.14	0.82	0.96	18.6
RP	#RP:32	End	65.75	-0.07	0.00	-0.00	-20.07	-0.26	0.0	-20.75	-0.69	-0.00	0.13	0.87	1.01	19.1
RP	#RP:32	Origin	65.75	-0.07	0.00	-0.00	-20.07	-0.26	0.0	-21.17	-0.69	-0.00	0.14	0.87	1.01	19.1
RP	RP:g	End	70.00	0.00	0.00	0.00	-23.01	-0.27	0.0	-21.17	-0.69	-0.00	0.13	0.91	1.04	19.4

Summary of Guy Tensions and Usages for Load Case "ASCE Design Example 4,I NA-"

Guy Label	Max. Tension (kips)	Allowable Tension (kips)	Factored Allowable (kips)	Usage %
G1	7.91	13.05	11.75	67.32
G2	11.64	13.05	11.75	99.14

Summary of Cable Tensions and Usages for Load Case "ASCE Design Example 4,I NA-"

Cable Label	Max. Tension (kips)	Allowable Tension (kips)	Factored Allowable (kips)	Usage %
TG1	3.57	13.05	11.74	30.42
TG2	4.12	13.05	11.74	35.07
TG3	1.52	13.05	11.75	12.93
TG4	2.31	13.05	11.75	19.67

DESIGN PARAMETERS:	TRANSMISSION POLE DESIGN EXAMPLE 4
80` C3 DOUGLAS FIR POLES WEIGHT OF WOOD 60 PCF MOR = 8,000 PSI Strength Factor = 0.671 MOE = 2,380,000 PSI	ASCE WOOD POLE STRUCTURE MANUAL OF PRACTICE

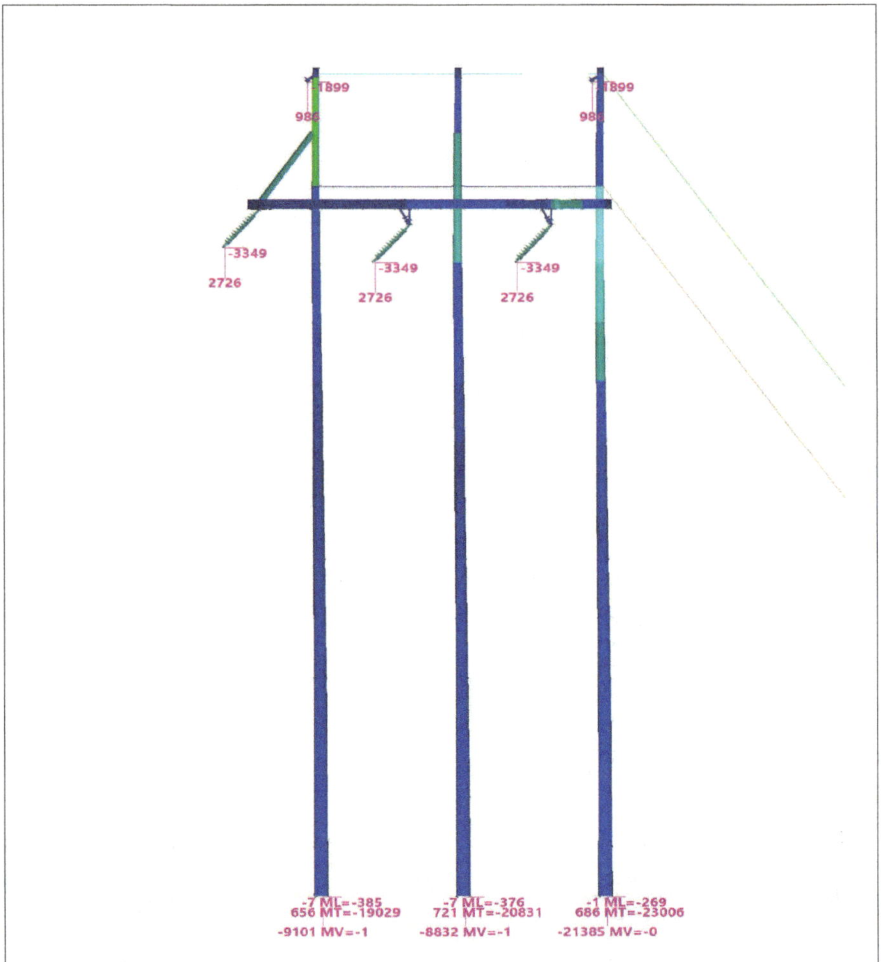

DESIGN PARAMETERS:	TRANSMISSION POLE DESIGN EXAMPLE 4
80` C3 DOUGLAS FIR POLES WEIGHT OF WOOD 60 PCF MOR = 8,000 PSI Strength Factor = 0.671 MOE = 2,380,000 PSI	ASCE WOOD POLE STRUCTURE MANUAL OF PRACTICE

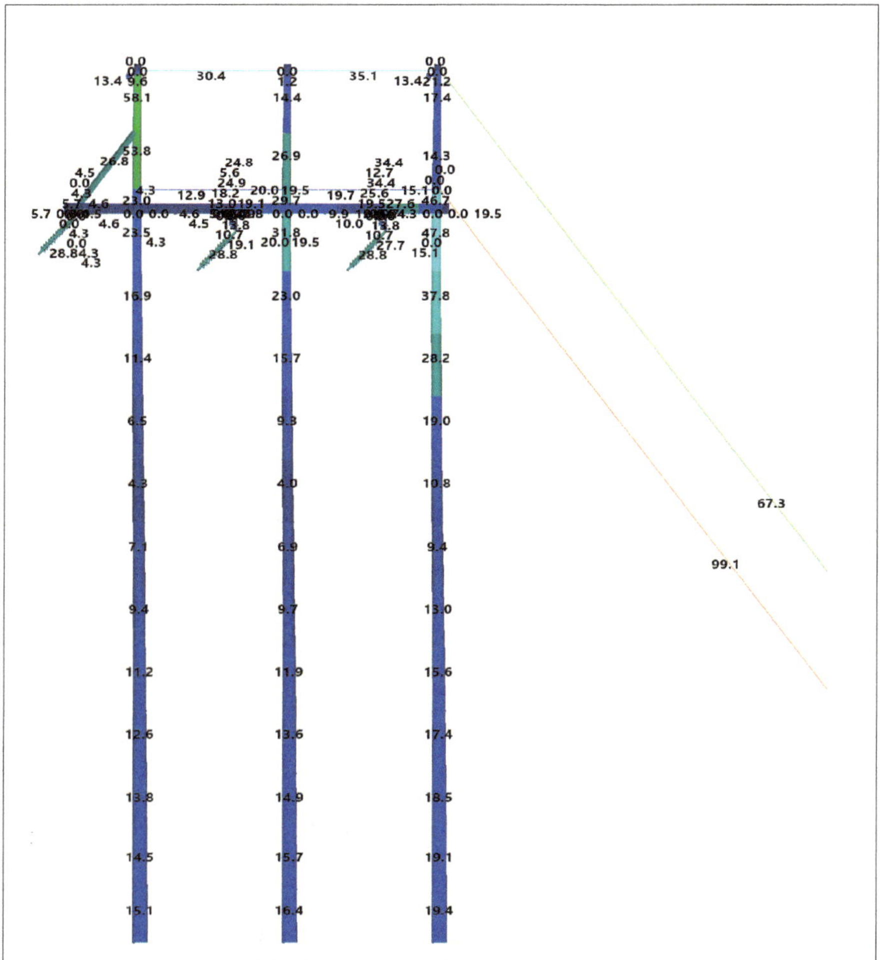

DESIGN PARAMETERS:
80` C3
DOUGLAS FIR POLES
WEIGHT OF WOOD 60 PCF
MOR = 8,000 PSI
Strength Factor = 0.671
MOE = 2,380,000 PSI

TRANSMISSION POLE
DESIGN EXAMPLE 4

ASCE WOOD POLE STRUCTURE
MANUAL OF PRACTICE

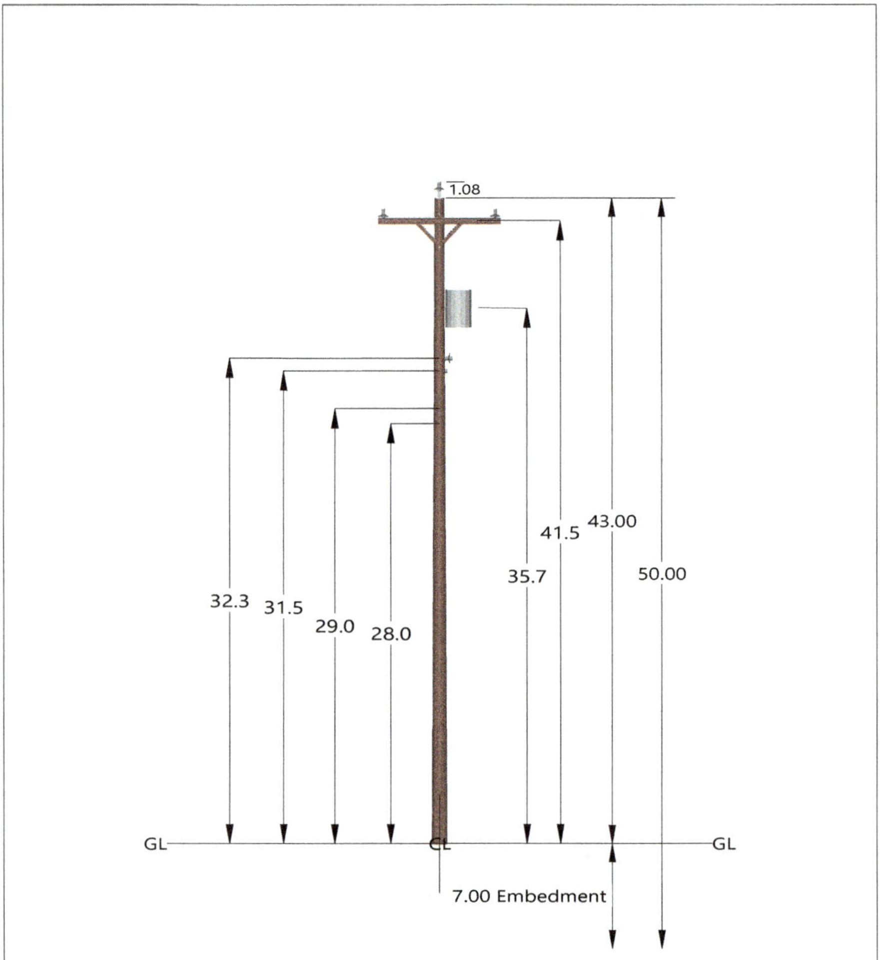

DESIGN PARAMETERS:	DISTRIBUTION POLE DESIGN EXAMPLE 5
50` C3 SOUTHERN PINE POLE WEIGHT OF WOOD 60 PCF MOR = 8,000 PSI Strength Factor = 0.671 MOE = 2,130,000 PSI	ASCE WOOD POLE STRUCTURE MANUAL OF PRACTICE

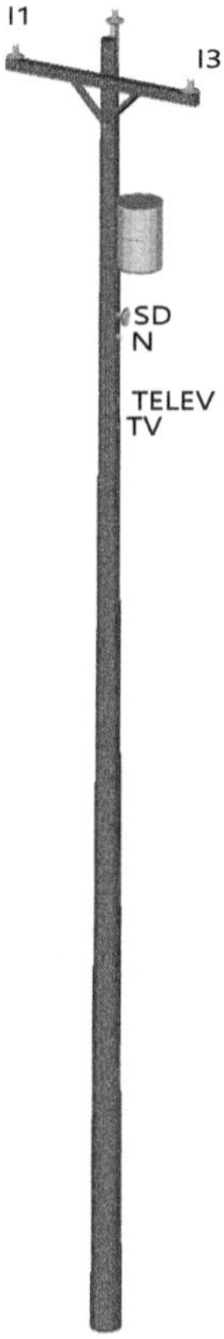

Loading Tree						
Load Case	Joint Label	Vert.	Trans. Load (lbs)	Long.	Trans. Wind (psf)
ASCE Example 5,I NA+	I1	553	208	0	7
	I2	553	208	0	7
	I3	553	208	0	7
	N	542	208	0	7
	SD	126	285	54	7
	TELEV	655	213	0	7
	TV	655	213	0	7

TRANSFORMER
WEIGHT - 250 LBS
WIND AREA - 3 FT^2
DRAG COEFF. - 1

DESIGN PARAMETERS:

50` C3
SOUTHERN PINE POLE
WEIGHT OF WOOD 60 PCF
MOR = 8,000 PSI
Strength Factor = 0.671
MOE = 2,130,000 PSI

DISTRIBUTION POLE DESIGN EXAMPLE 5

ASCE WOOD POLE STRUCTURE MANUAL OF PRACTICE

Detailed Wood Pole Usages for Load Case "ASCE Example 5,I NA+"

Element Label	Joint Label	Joint Position	Rel. Dist. (ft)	Trans. Defl. (in)	Long. Defl. (in)	Vert. Defl. (in)	Trans. Mom. (ft-k)	Long. Mom. (ft-k)	Tors. Mom. (ft-k)	Axial Force (kips)	Tran. Shear (kips)	Long. Shear (kips)	P/A (ksi)	M/S. (ksi)	Res. (ksi)	Max. Usage %
P	P:t	Origin	0.00	53.21	1.82	-3.56	-0.00	0.00	0.0	-0.01	0.00	-0.00	0.00	0.00	0.00	0.0
P	P:PTP1	End	0.33	52.52	1.79	-3.50	0.00	-0.00	0.0	-0.01	0.00	-0.00	0.00	0.00	0.00	0.0
P	P:PTP1	Origin	0.33	52.52	1.79	-3.50	0.00	0.33	-0.0	-0.17	0.95	-0.99	0.00	0.12	0.13	2.3
P	P:PTP2	End	1.00	51.13	1.74	-3.38	0.64	-0.33	-0.0	-0.17	0.95	-0.99	0.00	0.22	0.23	4.2
P	P:PTP2	Origin	1.00	51.13	1.74	-3.38	0.64	0.20	-0.1	-0.57	0.32	-0.00	0.01	0.20	0.22	4.0
P	P:ARM	End	1.50	50.10	1.71	-3.29	0.80	0.19	-0.1	-0.57	0.32	-0.00	0.01	0.24	0.26	4.8
P	P:ARM	Origin	1.50	50.10	1.71	-3.29	0.80	0.40	0.3	0.70	1.23	0.03	-0.02	0.27	0.29	5.3
P	P:G	End	2.25	48.54	1.66	-3.15	1.72	0.42	0.3	0.70	1.23	0.03	-0.02	0.50	0.51	9.6
P	P:G	Origin	2.25	48.54	1.66	-3.15	1.72	0.42	0.3	0.67	1.24	0.03	-0.01	0.50	0.51	9.5
P	P:B	End	3.15	46.68	1.60	-2.99	2.84	0.45	0.3	0.67	1.24	0.03	-0.02	0.78	0.79	14.8
P	P:B	Origin	3.15	46.68	1.60	-2.99	2.84	-0.35	0.2	-1.79	0.97	-0.01	0.04	0.77	0.81	15.2
P	#P:0	End	5.15	42.60	1.46	-2.64	4.78	-0.37	0.2	-1.79	0.97	-0.01	0.04	1.18	1.22	22.7
P	#P:0	Origin	5.15	42.60	1.46	-2.64	4.78	-0.37	0.2	-1.85	0.98	-0.01	0.04	1.18	1.22	22.7
P	#P:1	End	6.24	40.40	1.38	-2.45	5.85	-0.39	0.2	-1.85	0.98	-0.01	0.04	1.38	1.41	26.3
P	#P:1	Origin	6.24	40.40	1.38	-2.45	5.85	-0.39	0.2	-1.89	0.99	-0.01	0.04	1.38	1.42	26.4
P	P:XFMR	End	7.33	38.23	1.31	-2.27	6.94	-0.40	0.2	-1.89	0.99	-0.01	0.04	1.87	1.91	35.7
P	P:XFMR	Origin	7.33	38.23	1.31	-2.27	7.53	-0.40	0.2	-2.42	1.10	-0.02	0.05	2.03	2.09	38.9
P	#P:2	End	9.00	34.98	1.20	-2.01	9.37	-0.43	0.2	-2.42	1.10	-0.02	0.04	1.96	2.00	37.3
P	#P:2	Origin	9.00	34.98	1.20	-2.01	9.37	-0.43	0.2	-2.49	1.11	-0.02	0.05	1.96	2.01	37.4
P	P:SD	End	10.67	31.81	1.09	-1.75	11.21	-0.45	0.2	-2.49	1.11	-0.02	0.05	2.61	2.66	49.6
P	P:SD	Origin	10.67	31.81	1.09	-1.75	11.26	-0.46	0.1	-2.63	1.42	-0.07	0.05	2.62	2.67	49.8
P	P:N	End	11.50	30.27	1.03	-1.63	12.44	-0.51	0.1	-2.63	1.42	-0.07	0.05	2.71	2.76	51.5
P	P:N	Origin	11.50	30.27	1.03	-1.63	12.70	-0.51	0.1	-3.19	1.71	-0.07	0.06	2.77	2.83	52.8
P	#P:3	End	12.75	27.99	0.96	-1.46	14.84	-0.60	0.1	-3.19	1.71	-0.07	0.05	2.67	2.72	50.6
P	#P:3	Origin	12.75	27.99	0.96	-1.46	14.84	-0.60	0.1	-3.26	1.71	-0.07	0.05	2.67	2.72	50.7
P	P:TELE	End	14.00	25.78	0.88	-1.29	16.97	-0.69	0.1	-3.26	1.71	-0.07	0.06	3.34	3.40	63.3
P	P:TELE	Origin	14.00	25.78	0.88	-1.29	17.22	-0.70	0.1	-3.94	2.01	-0.08	0.07	3.39	3.46	64.4
P	P:TV	End	15.00	24.07	0.82	-1.17	19.24	-0.77	0.1	-3.94	2.01	-0.08	0.07	3.63	3.70	69.0
P	P:TV	Origin	15.00	24.07	0.82	-1.17	19.49	-0.77	0.1	-4.65	2.31	-0.08	0.08	3.68	3.76	70.1
P	#P:4	End	17.00	20.77	0.71	-0.94	24.11	-0.93	0.1	-4.65	2.31	-0.08	0.07	3.68	3.75	69.8
P	#P:4	Origin	17.00	20.77	0.71	-0.94	24.11	-0.93	0.1	-4.77	2.29	-0.08	0.07	3.68	3.75	69.9
P	#P:5	End	19.00	17.69	0.60	-0.74	28.69	-1.08	0.1	-4.77	2.29	-0.08	0.07	4.07	4.13	77.0
P	#P:5	Origin	19.00	17.69	0.60	-0.74	28.69	-1.08	0.1	-4.91	2.27	-0.08	0.07	4.07	4.14	77.1
P	#P:6	End	21.00	14.84	0.51	-0.57	33.23	-1.23	0.1	-4.91	2.27	-0.08	0.07	4.38	4.45	82.9
P	#P:6	Origin	21.00	14.84	0.51	-0.57	33.23	-1.23	0.1	-5.05	2.25	-0.07	0.07	4.38	4.45	82.9
P	#P:7	End	23.00	12.23	0.42	-0.43	37.72	-1.38	0.1	-5.05	2.25	-0.07	0.06	4.64	4.70	87.6
P	#P:7	Origin	23.00	12.23	0.42	-0.43	37.72	-1.38	0.1	-5.19	2.22	-0.07	0.07	4.64	4.70	87.6
P	#P:8	End	25.00	9.86	0.34	-0.31	42.16	-1.52	0.1	-5.19	2.22	-0.07	0.06	4.84	4.90	91.3
P	#P:8	Origin	25.00	9.86	0.34	-0.31	42.16	-1.52	0.1	-5.35	2.18	-0.07	0.07	4.84	4.90	91.3
P	#P:9	End	27.00	7.76	0.26	-0.22	46.52	-1.66	0.1	-5.35	2.18	-0.07	0.06	4.99	5.05	94.2
P	#P:9	Origin	27.00	7.76	0.26	-0.22	46.52	-1.66	0.1	-5.50	2.15	-0.07	0.06	4.99	5.06	94.2
P	#P:10	End	29.00	5.90	0.20	-0.14	50.82	-1.80	0.1	-5.50	2.15	-0.07	0.06	5.11	5.17	96.3
P	#P:10	Origin	29.00	5.90	0.20	-0.14	50.82	-1.80	0.1	-5.67	2.11	-0.07	0.06	5.11	5.17	96.3
P	#P:11	End	31.00	4.31	0.15	-0.09	55.04	-1.93	0.1	-5.67	2.11	-0.07	0.06	5.19	5.25	97.7
P	#P:11	Origin	31.00	4.31	0.15	-0.09	55.04	-1.93	0.1	-5.84	2.07	-0.07	0.06	5.19	5.25	97.8
P	#P:12	End	33.00	2.97	0.10	-0.05	59.19	-2.06	0.1	-5.84	2.07	-0.07	0.06	5.23	5.29	98.6
P	#P:12	Origin	33.00	2.97	0.10	-0.05	59.19	-2.06	0.1	-6.01	2.03	-0.06	0.06	5.23	5.30	98.7
P	#P:13	End	35.00	1.89	0.06	-0.03	63.25	-2.18	0.1	-6.01	2.03	-0.06	0.06	5.26	5.32	99.1
P	#P:13	Origin	35.00	1.89	0.06	-0.03	63.25	-2.18	0.1	-6.20	1.99	-0.06	0.06	5.26	5.32	99.1
P	#P:14	End	37.00	1.06	0.04	-0.01	67.22	-2.30	0.1	-6.20	1.99	-0.06	0.06	5.26	5.32	99.1
P	#P:14	Origin	37.00	1.06	0.04	-0.01	67.22	-2.31	0.1	-6.38	1.94	-0.06	0.06	5.26	5.32	99.1
P	#P:15	End	39.00	0.47	0.02	-0.00	71.10	-2.42	0.1	-6.38	1.94	-0.06	0.06	5.24	5.30	98.8
P	#P:15	Origin	39.00	0.47	0.02	-0.00	71.10	-2.42	0.1	-6.58	1.89	-0.06	0.06	5.24	5.30	98.8
P	#P:16	End	41.00	0.12	0.00	-0.00	74.88	-2.54	0.1	-6.58	1.89	-0.06	0.06	5.21	5.27	98.2
P	#P:16	Origin	41.00	0.12	0.00	-0.00	74.88	-2.54	0.1	-6.78	1.84	-0.05	0.06	5.21	5.27	98.2
P	P:g	End	43.00	0.00	0.00	0.00	78.57	-2.64	0.1	-6.78	1.84	-0.05	0.06	5.17	5.22	97.3

DESIGN PARAMETERS:	DISTRIBUTION POLE DESIGN EXAMPLE 5
50` C3 SOUTHERN PINE POLE WEIGHT OF WOOD 60 PCF MOR = 8,000 PSI Strength Factor = 0.671 MOE = 2,130,000 PSI	ASCE WOOD POLE STRUCTURE MANUAL OF PRACTICE

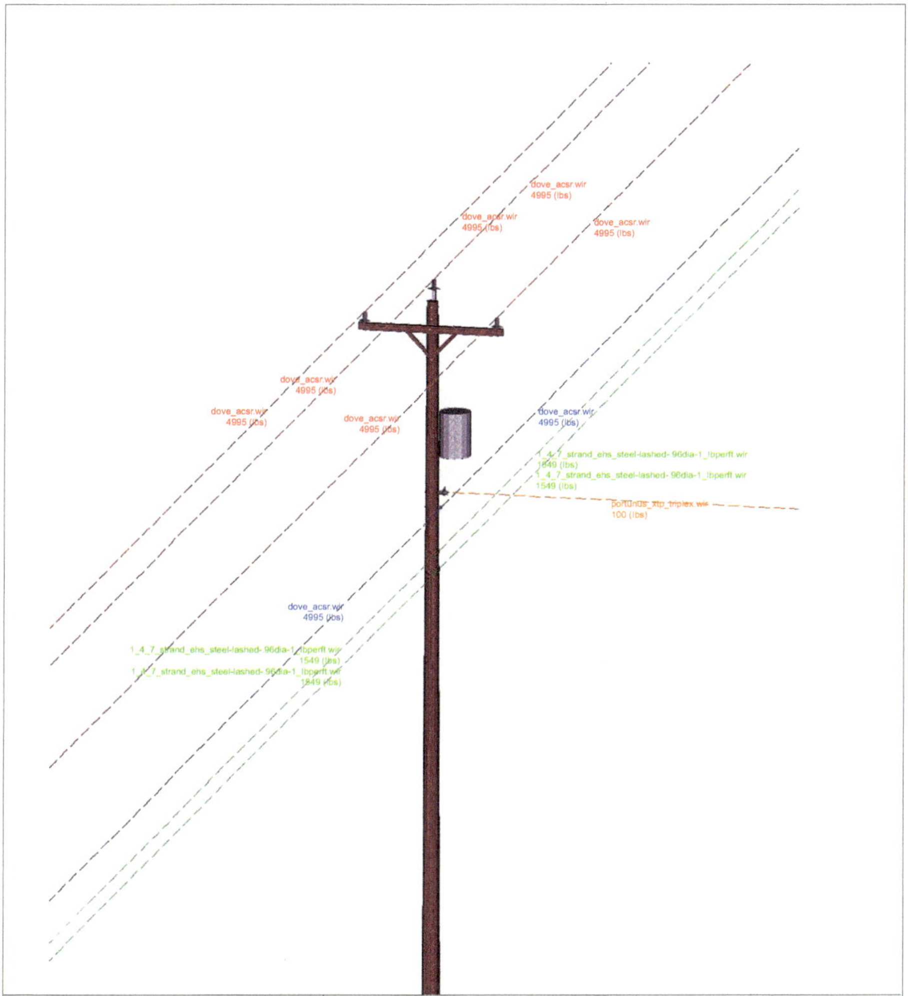

DESIGN PARAMETERS:	DISTRIBUTION POLE DESIGN EXAMPLE 5
50` C3 SOUTHERN PINE POLE WEIGHT OF WOOD 60 PCF MOR = 8,000 PSI Strength Factor = 0.671 MOE = 2,130,000 PSI	ASCE WOOD POLE STRUCTURE MANUAL OF PRACTICE

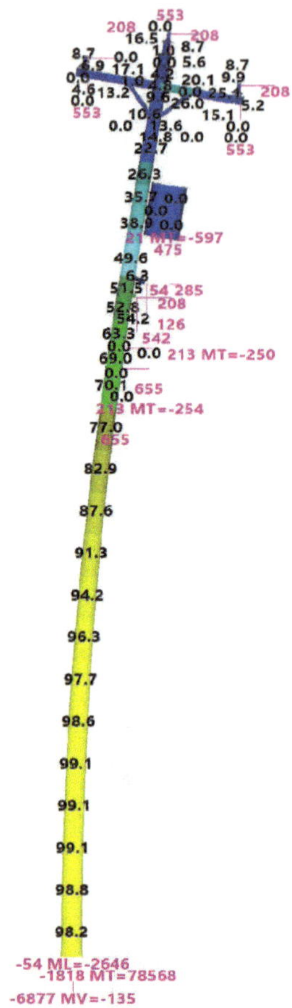

DESIGN PARAMETERS:	DISTRIBUTION POLE DESIGN EXAMPLE 5
50\` C3 SOUTHERN PINE POLE WEIGHT OF WOOD 60 PCF MOR = 8,000 PSI Strength Factor = 0.671 MOE = 2,130,000 PSI	ASCE WOOD POLE STRUCTURE MANUAL OF PRACTICE

APPENDIX C
LAMINATED WOOD POLES

C.1 INTRODUCTION

Laminated wood poles are a viable alternative to consider when round poles cannot meet the needs of a particular structure design. As an engineered product, laminated wood poles can be designed and fabricated to accommodate many loadings and scenarios as an alternative to round wood, steel, concrete, or fiber-reinforced polymer (FRP) poles.

C.2 APPLICATIONS

There are many engineering reasons for considering a laminated wood pole structure. They can be used for dead-end structures, angle structures, unbalanced loading conditions, and other self-supporting structures in situations where naturally grown round poles are inadequate. Laminated wood poles can also be used for specialty structures such as switches or other large equipment structures, substation structures, taller structures, and larger structures that round poles are not able to accommodate.

Laminated wood poles may be selected not only for engineering reasons but for a variety of other applications, including aesthetic purposes.

Other benefits of laminated wood poles that should be considered in their selection are lead times, natural resource sustainability, design performance, deflection limitations, aesthetics, field modifications, and ease of climbing. Laminated wood poles can be one-piece poles or multisection poles should that be a project requirement for shipping or other considerations.

C.3 GENERAL CONSIDERATIONS

As with round wood poles, the effect of impact and/or vibration loads sometimes observed on electrical transmission and distribution lines are generally reduced because of the energy-absorption characteristics of wood pole construction. But unlike round wood poles, laminated wood poles can be engineered and designed to meet a variety of needs, much like a steel, concrete, or FRP pole structure. The limitation to the variability in dimensions, the virtual elimination of defects and other anomalies, and the consistency of materials used reduce the coefficient of variation (COV) compared to round wood poles. ANSI O5.2, *Structural Glued Laminated Timber for Utility Structures* (ANSI/AWPA 2012), should be referenced in conjunction with this appendix when specifying and designing laminate wood pole structures.

Because laminated wood poles can be larger than round wood poles and are often rectangular in shape, consideration should be given to the foundation design and construction as the excavations can be significantly larger than those for other types of poles, both round wood and other materials such as steel, concrete and FRP.

C.4 STRUCTURAL DESIGN OBJECTIVE AND METHODOLOGY

Similar to round wood poles and other materials used in the industry, laminated wood poles should use Load and Resistance Factor Design (LRFD) methods, where the structure has a greater resistance (strength) capability than the maximum load and/or load effects expected to be experienced. This can be expressed in an LRFD format as

$$\emptyset R \geq \gamma Q \tag{C-1}$$

where

$\emptyset$ = Strength factor,

R = Measure of resistance (nominal strength),

γ = Load factor, and

Q = Calculated load or load effects.

For laminated wood poles, nominal strength is the engineered fiber strength, or modulus of rupture (MOR) in bending. The engineered fiber strength is provided by the manufacturer. As laminated wood poles are rectangular in nature, there will be an MOR in the transverse direction and a separate MOR for the longitudinal direction on the pole. These MOR values shall be determined by the laminated wood pole manufacturer using a 5% lower exclusion limit (LEL_5).

Similarly, the laminated wood pole manufacturer shall provide a corresponding modulus of elasticity (MOE) for the designed wood poles.

C.4.1 Analysis and Design

As with round wood poles, the primary consideration is to ensure that the applied loads (with appropriate load factors) do not exceed the pole strength as calculated with applied strength reduction factors.

Probabilistic loads and load factors are published in *Guidelines for Electrical Transmission Line Structural Loading* (ASCE 2010).

Deterministic loads—such as the National Electric Safety Code (IEEE 2017) and California's General Order 95 (California PUC 2018)—should use the load and strength factors dictated by those standards.

Also, as with round wood poles and other structural materials, nonlinear or "P-delta" methods should be used in the analysis and design of all laminated wood pole structures. This analysis is best accomplished by computer analysis software that can provide a sophisticated and accurate structural analysis. The reactions at various sections along the pole should be calculated using the stress equations presented in the next section, C.4.2.

C.4.2 Stress Equations

Laminated wood poles used as transmission or distribution line structures will be subject to the three primary stresses: bending, axial, and shear stresses. The bending stress and axial stress are generally significant enough to warrant inclusion in an analysis. However, the stress associated with shear loads (perpendicular to the grain) are generally small compared to the other loads and are often neglected.

As with round wood poles, the basic formula for bending stress is

$$f_b = M_r / S \tag{C-2}$$

where

f_b = Fiber stress in bending,

M_r = Sum of all applied primary and secondary bending moments acting on the cross-section being analyzed, and

S = Section modulus of the pole cross-section being analyzed.

As laminated wood poles are usually rectangular in cross-section, there may be a separate bending stress for the transverse direction (f_{bt}) and the longitudinal directions (f_{bl}). The basic formula for axial stress is

$$f_a = P / A \tag{C-3}$$

where

f_a = Stress on the section owing to axial load,

P = Sum of all axial loads acting on the cross-section being analyzed, and

A = Cross-sectional area of the cross-section being analyzed.

Thus, the combined bending and axial stress equation for laminated wood poles is

$$(f_{bt} / F_{bt}) + (f_{bl} / F_{bl}) + f_a / F_a \leq 1.0 \quad \text{(C-4)}$$

where
F_{bt} = Allowable transverse bending stress,
F_{bl} = Allowable longitudinal bending stress, and
Fa = Allowable axial stress.

C.4.3 Defects

Two kinds of defects can reduce the strength of a laminated wood pole: bolt holes and section loss due to damage.

A bolt hole going through the pole in either the transverse or longitudinal direction will reduce the overall axial and bending capacity of the pole. The bolt hole will reduce the overall area as well as the section modulus as formulated by the following equations.

Reduction in cross-section area:

$$b \times \text{length of hole} \quad \text{(C-5)}$$

Reduction in transverse section modulus:

$$S_t = bw^2/6 \text{ for a bolt hole in the transverse direction} \quad \text{(C-6)}$$

Reduction in transverse section modulus:

$$S_t = tb^2/6 \text{ for a bolt hole in the longitudinal direction} \quad \text{(C-7)}$$

Reduction in longitudinal section modulus:

$$S_l = wb^2/6 \text{ for a bolt hole in the transverse direction} \quad \text{(C-8)}$$

Reduction in longitudinal section modulus:

$$S_l = bt^2/6 \text{ for a bolt hole in the longitudinal direction} \quad \text{(C-9)}$$

where
w = Transverse width of the pole,
t = Longitudinal width of the pole, and
b = Hole diameter.

Loss of axial and bending strength at a location on a pole because of natural and human-created damage can occur. For example, a defect could be characterized by $\alpha T = 0.6$ and $\alpha L = 0.8$, meaning that the pole transverse bending capacity is only 60% of what it would be without the defect, and its longitudinal bending capacity is only 80% of what it would be without the defect. To account for the effect of αT and αL, appropriate strengths can be multiplied by their respective axial and bending-capacity reductions.

C.4.4 Foundations

There are several commonly used foundations for laminated wood poles.

C.4.4.1 Direct Embedded. As with round wood, the most common foundation for laminated wood poles is to direct-embed them into the soil (see Figure C-1). The engineer can either specify the embedment depth or provide soils information for the manufacturer to design the foundation system with the applied loads.

Figure C-1. Direct-embedded laminated wood pole.
Source: Jaron Reay, reprinted with permission.

Figure C-2. Laminated dead-end structure on a concrete pier foundation. Source: Kishor Kumar, reprinted with permission.

C4.4.2 Steel Base Connection. In some cases, such as dead-end locations, it may be desired to mount a laminated wood pole to a concrete caisson foundation with a special steel base connection. The steel base connection should be able to handle biaxial moment combined with axial forces.

A steel base connection should be designed as part of structure components by the laminated wood pole supplier. A typical steel base connection is presented in Figures C-2 and C-3 for illustrative purposes.

Figure C-3. Typical steel base connection.
Source: Kishor Kumar, reprinted with permission.

REFERENCES

ANSI/AWPA. 2012. ANSI O5.2, *Structural glued laminated timber for utility structures*. Birmingham, AL: American Wood Protection Association.

ASCE. 2010. *Guidelines for electrical transmission line structural loading*, MOP 74. Reston, VA: ASCE.

California PUC (Public Utilities Commission). 2018. *Rules for overhead line construction*, General Order No. 95. Accessed November 18, 2018. http://www.cpuc.ca.gov/gos/GO95/go_95_startup_page.html.

IEEE (Institute of Electrical and Electronic Engineers). 2017. ANSI Standard C2, *National electrical safety code*. New York: IEEE.

APPENDIX D

QUALITY AND STRENGTH ASSESSMENT TOOLS AND DEVICES

D.1 INTRODUCTION

Several nondestructive evaluation (NDE) or nondestructive testing (NDT) inspection devices are available to provide information on pole condition and/or strength. (*Note:* Although the terms *NDE* and *NDT* have different technical implications, they are often used interchangeably, hence the term *NDE* is used exclusively in the following narrative.) Some NDE devices are intended to replace or enhance conventional inspection techniques (Morell 2000, CEATI 2001); others are intended to provide information that cannot be determined through conventional means of inspection. Visual plus sound-and-bore inspection (with excavation as appropriate) is the most widely used methodology for wood pole inspection and assessment. Continued enhancements to NDE tools and devices may serve to enhance the decision-making ability of the inspector or aid in the inspection and assessment process.

Inherently, NDE tools and conventional processes have accuracy limitations in their estimates of pole condition and/or strength. Well-trained inspectors can minimize the likelihood of significant errors through their experience and knowledge of how to apply inspection processes and technologies.

The degree to which NDE technologies are adopted is a function of whether they improve the value or lower the cost of traditional inspection equipment and processes. Unless an NDE technology improves the accuracy of inspection results over what can be achieved via a visual plus sound-and-bore inspection, it will have a low probability of acceptance—especially since excavation and boring will still be required for application of remedial treatments. However, wood pole owners and inspection contractors both have an interest in NDE methods to potentially reduce labor efforts, using the device to determine pole condition and strength rather than the inspector's assessment.

With any NDE technology, it is important to remain mindful of the difference between what is being measured and what is inferred from that measurement. This is particularly true for strength assessment tools because differences among the growth properties of various pole species, moisture content variations, and conditions such as radial checking or separation between growth rings may alter NDE parameters while having little effect on the ultimate strength of the pole.

D.2 NONDESTRUCTIVE EVALUATION SYSTEMS

When evaluating the potential application of an NDE technique, it is important to recognize that there are two aspects to assessing the maintainability of wood pole structures: quality and strength. Quality is a function of physical condition, and strength is the capacity to support the design loads. The following sections describe the potential applications and limitations of commonly available NDE technologies.

D.2.1 Ultrasonic Tomography

The technology is based on the transmission of ultrasonic signals and measuring the velocity of the signals received at typically four to six sonic transducers (i.e., receivers) located at regular intervals around the circumference of the pole. The ultrasonic velocities are then used by a mathematical algorithm to generate a predicted image of this planar cross-section. The resulting cross-sectional image, remaining section modulus, and area are produced off-line after the mathematical processing.

D.2.2 Sonic/Acoustic

The technology typically represents the introduction of kinetic energy into the wood component by means of a hammer or pendulum device. The resulting sonic waves are recorded by a sensor located in the same plane as the impact point of the hammer or pendulum. Measurements are typically captured and stored electronically for use in modeling by proprietary software applications. Assessment of mechanical properties such as a pole's bending strength are projected through empirical models based on destructive testing.

D.2.3 Radiographic Imaging and Ground-Penetrating Radar (GPR)

The base technologies of radiographic imaging and GPR are relatively similar as they each rely on the proper modeling of reflected waves from a supply and reflective surface. Similar to sonic/ultrasonic technologies, some radiographic technologies can be used in either scanning mode to assess pole condition or in a decay/damage mapping mode to create an image of the condition of the cross-section of the pole. Proprietary software analyzes the signal response and provides modeling of mechanical properties such as remaining section modulus, moment of inertia, and area.

D.2.4 Resistance Drilling

Resistance drilling technology is used in wood component inspection and is somewhat similar to sound-and-bore practices used in conventional wood pole inspection. Resistance or torque measurements required to maintain a constant velocity of drill-bit movement are recorded. A drop in the resistance measurement indicates a change in wood density, presumably because of some form of degradation (e.g., fungal decay or insect attack), with sudden drops being indicative of low-density material or voids. A supplemental software application is necessary to process the data collected in order to estimate remaining pole strength.

D.2.5 Mechanical Penetration Resistance/Electrical Resistance Measurement

The base technology was developed to assess pole condition and/or strength using a measurement of the force required to push a probe into a wood pole combined with the electrical resistance measured between the probes. The force required to push probes into the surface of a pole can provide some indication of wood density and fiber strength. The models are dependent on relatively large databases of destructive test results relative to measured NDE parameters for the particular species being evaluated.

D.2.6 Load-Deflection Measurement

Technologies that assess the remaining strength of poles based on deflection measurements made while applying a load to the pole rely on the relationship between the modulus of rupture (MOR) and the modulus of elasticity (MOE) to predict the ultimate strength. While there is a correlation between MOR and MOE, there is quite a bit of variability in the correlation, and research has shown that the quality of the correlation decreases as poles degrade.

REFERENCES

CEATI International (Centre for Energy Advancement through Technological Innovation). 2001. "Validating NDE tools for evaluating present condition, residual strength, and remaining life of in-service wood poles." *Tech. Rep. No. T004700-5005*. Montreal, QC: CEATI.

Morrell, J. J., and R. G. Rhatigan. 2000. "Pole inspection: What do the devices tell you?" *Proc., Canadian Wood Preservation Association, November*. Vancouver, BC: CWPA, 31–44.

APPENDIX E
FOUNDATION EXAMPLE

This appendix takes an example from Appendix B and uses two common methods—the Brinch Hansen method and the modified Broms' method—to calculate the required embedment depth for a typical wood pole structure. Table E-1 shows the calculated embedment depth for a 75 ft class H1 Douglas-fir wood pole in six different soil types using both methods and compares them to the rule-of-thumb embedment depth.

The six soil types are

1. Dense dry sand,
2. Medium dry sand,
3. Loose dry sand,
4. Stiff saturated clay,
5. Medium saturated clay, and
6. Soft saturated clay.

Table E-1. Wood pole embedment comparison.

Wood Pole	ANSI Pole Diameter at Groundline, ft	Soil Type	Density Υ kip/cft	Friction Angle or Shear Strength (f/ft^2)	Loads at Groundline		Pole Embedment, ft		
					Shear, kips	Moment, ft-k	Rule of Thumb (10%+2ft)	Brinch Hansen Method	Modified Broms'
DF-75-H1	1.437	Dense Dry Sand	0.140	41	4.1	214	9.5	9	10
DF-75-H1	1.437	Medium Dry Sand	0.120	33	4.1	214	9.5	11.5	11.5
DF-75-H1	1.437	Loose Dry Sand	0.095	28	4.1	214	9.5	14	13.5
DF-75-H1	1.437	Stiff Saturated Clay	0.140	2	4.1	214	9.5	8	6.5
DF-75-H1	1.437	Medium Saturated Clay	0.120	0.75	4.1	214	9.5	13	11
DF-75-H1	1.437	Soft Saturated Clay	0.100	0.25	4.1	214	9.5	22.5	20

Note: This example does not take deflection or rotation into consideration.

INDEX